Beyond the Bloom

Bloom refining and iron artifact production in the Roman world

David Sim

Edited by

Isabel Ridge

BAR International Series 725
1998

Published in 2016 by
BAR Publishing, Oxford

BAR International Series 725

Beyond the Bloom

ISBN 978 0 86054 901 7

Volume Editor: Rajka Makjanić

BAR Publishing is the trading name of British Archaeological Reports (Oxford) Ltd. British Archaeological Reports was first incorporated in 1974 to publish the BAR Series, International and British. In 1992 Hadrian Books Ltd became part of the BAR group. This volume was originally published by Archaeopress in conjunction with British Archaeological Reports (Oxford) Ltd / Hadrian Books Ltd, the Series principal publisher, in 1998. This present volume is published by BAR Publishing, 2016.

Printed in England

BAR
PUBLISHING

BAR titles are available from:

BAR Publishing
122 Banbury Rd, Oxford, OX2 7BP, UK
EMAIL info@barpublishing.com
PHONE +44 (0)1865 310431
FAX +44 (0)1865 316916
www.barpublishing.com

It may be as many sing,
Gold is over all men king
Yet if truth be told
'Tis with iron men win gold.

To John Anstee a good friend

Acknowledgements

I would like to thank Dr. Isabel Ridge for her unstinting help, sound guidance and unwavering support, with out her this book would have taken even more toil and tears than it did. Dr. Jamey Kaminski and Hugo Mills for all their help in solving the many computing problems. Professors Michael Fulford and John Allan who supervised my PhD and together with Dr Paul Craddock encouraged me to write this book. Mr Charles Honor who acted as my striker. My very good friend Mr John Anstee whose work inspired me to take up experimental archaeology in the first place. Professor Richard Chaplin for his continuing encouragement and support.

Foreword

The research for "Beyond the Bloom" was carried out between 1991 and 1994 and was submitted as my PhD thesis, which forms the basis of this book. I am very grateful to Dr. Isabel Ridge who has edited it for publication. I have taken the opportunity to clarify some points and add new material not included in the original thesis.

The basic philosophy that underlies this book is that the best way to understand a complex system is to try to reproduce it. The book is concerned with the technological processes that are involved in transforming raw iron made by the bloomery process into finished artifacts.

The major part of this work details practical experiments that replicate the working environment of a Roman blacksmith. The tools and equipment used were as far as possible copies of Roman originals. A record was kept of the time taken to turn raw bloom iron into workable iron and the amount of fuel and other materials consumed. Similar records were made of the times to make Roman iron artifacts together with the amount of metal and fuel consumed.

The distribution of hammer scale on the floor around the anvil was investigated to establish if a pattern exists that can be compared with the distribution of hammer scale found in the archaeological record. The origin of hammer scale produced by experiments is a known quantity, therefore morphological and chemical analysis has been carried out so that the results can be compared with hammer scale found in the archaeological record.

Several modern forges were visited and samples were taken from the floors. These contained hammer scale and other materials used in those forges. Theses samples were used for comparison with the products of experiments.

The work provides a data base that will enable archaeologists to compare evidence found in the archaeological record with the experimental results.

David Sim

Reading
September 1998

CONTENTS

LIST OF TABLES

Chapter 3

Chapter 4

LIST OF FIGURES

Chapter 2

Chapter 3

Chapter 5

CHAPTER 1

The place of iron in the Roman world

'Without iron human life is simply impossible.'

Thomas More, Utopia.

Introduction

Steel is present in almost every aspect of modern daily life. It is used in everything from skyscrapers to nails in a fence, from space travel to roller skates. The world we inhabit today would be impossible without steel just as the world inhabited by the Romans would have been impossible without iron.

Iron items from the archaeological record seldom arouse great interest. To look at they are often no more than a mud-like mass of iron oxide (rust) which give only faint hint of the time and level of workmanship invested in them. Iron artifacts are, however, the key to understanding a civilisation which the material affected in so many ways.

There are two main questions which need to be addressed at this point:

(i) what was iron used for; and,

(ii) how much iron was used.

Having considered these points we can move on to discuss:

(iii) the form of ferrous metals used in the Roman world;

(iv) information provided by the archaeological record on Roman iron; and,

(v) iron manufacturing processes: pre-industrial and industrial.

The object of this chapter is to present an overall view of the methods of iron smelting that were practised in the recent past by pre-industrial societies. It is not intended as an in-depth study of specific locations or peoples, but to describe the general methods that have been used to make iron.

The objective of iron making is to extract iron from its ore, and the process follows some general rules, which are: Mining produces iron ore which is then smelted; the product of smelting is bloom-iron; and bloom-iron is refined by heating and hammering to produce wrought iron which is then forged into artifacts.

1.1 Historical setting

Archaeological evidence suggests that Rome became a city around 575 BC (Stobart 1987, 20). The early Romans learned much from the Etruscans (Stobart 1987, 20) who were an iron-using civilisation in 1000 BC (Tylecote 1976, 46). It is a reasonable assumption that by the foundation of Rome, the technology of iron was well understood and practised in that part of the world. By the time of Rome's establishment, the use of most of the then-known metals was well established, and the expertise of extraction and processing was at an advanced level. It may be true to say that few inventions in iron technology occurred in the Roman period, but the uses of ferrous metal as well as the development (as opposed to invention) of technology made enormous strides. It seems very likely that the size, operational organisation, and level of technology of the Roman iron industry has so far been understated, but the lack of evidence makes it very difficult to establish an accurate model.

The use of iron in the Roman world

The use of iron in the Roman world was widespread and the metal played a vital role in almost all aspects of life. Manning (1985a) in the contents of his catalogue classified iron artifacts as follows:

(a) Metalworking tools
(b) Woodworking tools
(c) Sculptors', Masons' and Plasterers' tools
(d) Tools used for processing wool and cloth
(e) Leather work
(f) Agricultural tools
(g) Farriers' tools
(h) Transport
(j) Finger rings and bracelets
(k) Toilet implements
(l) Surgical instruments
(m) Shackles
(n) Styli
(o) Locks and keys
(p) Domestic Equipment
(q) Knives, razors and cleavers
(r) Structural fittings
(s) Miscellaneous
(t) Armour and military equipment
(u) Weapons

Some of these sections may be further sub-divided,

(a) Jewellery-making; Goldsmithing, Silversmithing, Copper and Bronzesmithing, and, Blacksmithing.
(b) The timber industry. Woodland management, Carpentry, Fine cabinet making.
(c) Quarrying, Stone dressing, Stone carving.
(e) Tanning, Cobbling.
(h) Land, Sea, River.
(r) Domestic, Military.

The above lists help to demonstrate the diverse applications in which iron was used.

It is true that most of the above activities were carried out before iron was available, but the introduction of iron tools was an improvement on the technology.

1.2 Volume of iron used in the Roman world

Iron has many advantages over other metals in the Roman period, significantly:

(i) It can be shaped by forging, an economical and quick processing technique compared with casting or shaping by metal removal with files etc; and,
(ii) Steel (the alloy of iron and carbon) tools can be made to hold an edge that will outlast a similar tool made of bronze.

Probably the most important property of iron was that by alloying it with carbon to make steel, which could be made harder than any then-known metal, tools could be made to cut other metals. Estimating the amount of iron used by the Romans is almost impossible. There are items that survive but there is no way of knowing what proportion these items represent of the total number of artifacts in circulation at any one time.

Some indication might be gained from the size of the residual slag heaps from Roman iron making sites; however so many Roman slag heaps were exploited as a source of metalling for roads or, because the slag was very rich in iron, it was re-smelted. This makes an accurate estimate of the original slag heap size very difficult.

The civil and military population had a continual need for iron simply because iron implements wear out in general use; every time a knife is sharpened some of the metal is lost and hinges for doors wear out because of metal abrading on metal. All iron items wear away at different rates and it is almost impossible to state for certain how long an item will last.

The widespread use of iron suggests that though it was essential to sustain life it was not held in high esteem by the Romans. It is not stretching the imagination too far to say that the Romans' view of iron was the same as modern man's view of steel, it was there but no one noticed it.

1.3 Ferrous metals of the Roman world

The word iron is used to describe a number of different metallic alloys that have iron (Fe) as their major constituent. Steel, cast iron, and wrought iron are often referred to by the umbrella term "iron". In view of this range of possibilities it is necessary to define the term iron as it is applied when considering the Roman period. In modern usage, iron is a term used to describe cast iron, which is an alloy of iron and 2% to 4% carbon. Cast iron was not deliberately produced in the Roman period, and any cast iron found in a Roman context is thought to be the result of accidents in smelting (Tylecote 1976, 55). The two ferrous metals in use in the Roman period and until the start of the 15th century were steel, an alloy of iron and carbon where the carbon content is less than 1.7%, and wrought iron, which is a mixture of iron and slag, with a carbon content in the range of about 0.02% to 0.15%.

The most common ferrous material used in Roman times was the product of the bloomery process - a metal that we now call wrought iron. The production of iron started with iron ore which was treated to extract the iron. Iron was then forged into artifact. These two parts of the continuum are certain, but the details of the intermediate processes are in considerable doubt.

Wrought iron in its purest form contains very little carbon - as little as 0.02% - which means that there is not sufficient carbon to allow it to be hardened by heating to red heat and quenching in oil or water, therefore hardening is achieved by other lengthy processes such as carburising (see glossary). The properties of a material are profoundly influenced not only by the microscopic impurities in the metal, but also by the presence of macroscopic impurities such as slag (silicates and oxides) which is characteristic of wrought iron. As wrought iron is worked the slag inclusions become elongated in the direction of working, which has an important effect on the properties of the metal. The slag is apparent in the transverse direction of working as round dark grey blobs and in the longitudinal section as thin streaks. Wrought iron is strongest along the direction of working and as further working proceeds the tensile strength increases. The tensile strength of wrought iron in the direction of working is 339 MN/m^2 to 401 MN/m^2 compared to 293 MN/m^2 to 355 MN/m^2 for modern mild steel. The elongation before fracture is similar in both materials. Good quality wrought iron is very tough and can be bent perpendicular to its axis either hot or cold, which makes it ideal for making such things as chains. This is where the direction of the slag inclusions often referred to as 'grain' is evident, as there is far more resistance to bending about an axis perpendicular to the grain than one parallel to it. If a bar of wrought iron is nicked with a hacksaw and bent it will fracture to reveal a fibrous structure. The fibres are grains of pure iron elongated in the direction of working with the slag embedded in them. The main impurities in wrought iron are carbon, sulphur, silicon and phosphorus.

The amount of carbon in wrought iron does not usually exceed 0.15% but it can rise to 0.25%. This will increase the tensile strength. With this high level of carbon, the metal will begin to resemble steel and will show signs of hardening when quenched from red heat. Sulphur may be injurious even in small amounts, rendering the iron brittle when it is either hot or cold. Silicon is always present partly in the slag and partly as silicate of iron dissolved in ferrite (see glossary). Phosphorus is always present and, as with silicon if dissolved in ferrite, will cause the metal to be brittle (Greaves and Wrighton 1957, 82).

The malleability of wrought iron is considerably increased when its temperature is raised to above 700°C. This makes it ideal for forging, but it can also be shaped when cold by means of metal removal, such as filing or grinding. Wrought iron can be made hard by carburization and then heat-treated in the same way as steel, but its toughness makes it ideal for weapons or tools that are going to be subjected to heavy shock forces such as might be encountered when using a hammer or a sword.

When the term iron is applied to artifacts from the Roman period it is most likely the material it refers to is wrought iron.

1.4 The archaeological record

The information concerning iron given up by the archaeological record can be divided into the following broad categories:

(a) raw residues (ore, fuel, furnace materials)
(b) the residues from smelting
(c) the residues from smithing
(d) artifacts

(a) Raw residues

On the whole, iron-making sites seem to be abandoned when the site is no longer viable and it is assumed that the site will be cleared and useful equipment will probably be taken away before the workforce moves on. This will mean that what remains are items that have become lost or covered up, immovable items such as furnaces and items that are of no further use such as slag. (Fulford and Allen 1992). The same will largely be true of smithing sites unless the site is overcome by a natural disaster such as the eruption of Vesuvius in AD. 79, which buried Pompeii. The blacksmiths' workshops within that city would reveal a very great deal but little research seems to have been done with the exception of Gralfs (1988, 104) who identified two blacksmiths and eight bronze working sites within the city.

(b) Residues from smelting

From smelting sites there will be furnace bottoms, some of the walls of the furnace (Tylecote 1962a, 19), perhaps the remains of iron ore and fuel and possibly clay. From these items it is possible to build a picture of some of the technology that was used on specific sites.

(c) Residues from smithing

Hammer scale and hearth bottoms are common items found during excavations of forges. The other finds from a blacksmith's workshop are usually material that has been lost. During the working life of the smithy, such as small iron off cuts and nails. When hammer scale is found it is usually in small quantities. Experiment conducted by the author showed that one blacksmith working alone produced approximately half a tonne of hammer scale per year (Ironstone Forge, Reading, see chapter 5). In theory hammer scale should be found in large quantities, but it is not. However it is possible that it was used in other processes. A certain amount of hammer scale mixed with sand is used by blacksmiths as a flux when fire-welding, but this will only account for a few kilograms a year. Hammer scale is a good abrasive and may have been used in the polishing of metal and other hard material. It is pure speculation, but it is possible that if hammer scale was used as a polishing agent, then this may account for the small amounts found when iron-working sites are investigated.

(d) Artifacts

From the moment that iron is separated from its ore and exposed to the atmosphere it starts a process of oxidisation that will return it to its natural state, that is, it starts to rust. This has a profound effect on iron objects found in the archaeological record.

Artifacts that are found in conditions of damp with a good supply of oxygen are very often only iron oxide in the shape of the original artifact, with, at times, a small core of iron left at the centre of the object. Others may be completely oxidised and all that is available for inspection is the shape of the artifact, and other than being a ferrous metal, little can be determined concerning its original composition.

Where un-oxidised iron remains, there are a number of tests which will provide information about the material properties. Standard metallurgical tests will give us tensile strength, elongation, impact resistance, hardness, etc. The optical microscope can be used to determine the structure of the metal. More elaborate tests are available such as using the scanning electron microscope (referred to as SEM), which will give very high resolution and a qualitative analysis of the elements present in the metal. X-Ray Fluorescence (referred to as XRF) will give a quantitative analysis, and radiocarbon dating using accelerator mass spectrometry can be used to give dates for iron (Cresswell 1991, 78).

All these tests are either destructive or intrusive. Tensile and impact tests break the sample, and so should only be used if there is enough material to allow some to be lost forever. Microscopic analysis needs a polished surface which will damage the specimen. SEM needs a small sample to be polished. For XRF, the sample is ground into a powder, as it is with radiocarbon dating (^{14}C). The use of destructive tests is very limited when dealing with ancient iron artifacts. If the sample is the only one of its kind or even one of a limited number in most cases any damaging test is out of the question. There will be times when large numbers of artifacts are found and then some parts can be sacrificed. Although it is true that much of the iron from the Roman period is in poor condition it is not so in every case. The Roman site of Vindolanda near Hadrian's Wall, and the site of Thorsburg in North Germany (Engelhardt 1866, 23-25), are both anaerobic and the iron finds from them are in almost perfect condition. Many items are in the condition in which they were when deposited. Inspection of these items has revealed much about the standard of workmanship of the blacksmiths who made them. The pattern-welded swords from Thorsburg (Engelhardt 1866) show a level of workmanship that can only be considered as first class, though it is not known if this collection of over 100 swords is the product of local craftsmen or Roman smiths. It is a reasonable assumption that the items found at Vindolanda are the product of Roman or Roman-trained blacksmiths and again, they demonstrate a high degree of skill and craftsmanship. Metallurgical examination of the swords has not been possible (due to restrictions at the museum where they are housed) and so the quality of iron cannot be commented on. The level of craftsmanship displayed in these items can leave no doubt that the standard of Roman iron working was as high as anything that followed in later periods.

1.5 Pre-industrial and industrial iron working

The original practitioners of smelting have long since disappeared, and we have no manuals describing their trade. In order to obtain some clues to the practices of the past, it has been customary to turn to societies that still use pre-industrial methods to produce iron. Until the 1960s there were still parts of Africa where iron was smelted by pre-industrial methods and it is to these that attention was turned.

A survey of pre-industrial iron-making was conducted by Tylecote (1965). He states that iron-making by older methods was then fast dying out, and the practice has now completely finished. In spite of diligent search, no trace has been found of primitive iron-making still being carried out as a method of supply of iron for consumption. Edwin Okafor (pers. comm. 1992) stated that in Africa there are no longer any tribes that produce iron by the pre-industrial method, but occasional smelts are performed for the benefit of archaeologists and anthropologists and it is from these that extrapolations are made. There are still blacksmiths who work iron artifacts by traditional African blacksmithing methods. The prime reason given for the disappearance of primitive iron making is the vast amount of scrap iron available from abandoned cars and the general detritus of metal from the colonial period that litters Africa.

The iron smelting technology of the Roman period can be described as fitting within the following parameters. All iron was produced by the direct method of iron production; deliberate fluxing of the smelt was not practised; and under the right conditions, the bloomery process could be used to produce steel.

The model for some experiments in early iron-making has been taken from pre-industrial societies. Pre-industrial is taken to mean non-mechanised and usually low volume output. Sadly for the archaeo-metallurgist, such peoples have all but disappeared.

There are two instances of iron making that warrant special detailing, they are the Japanese Tatara method of iron making and the Mafa iron smelting of North Cameroon, both described in chapter 2. A study of their methods shows how the common problem of turning bloom iron into workable iron has been solved in different ways.

Extraction and preparation of iron ore

The mining of iron ore can be very simple. It seems that in many cases in Africa the mining was open-cast and the ore was extracted from trenches or small pits. In some parts of Africa, India and Japan, coarse sand containing iron is washed to extract the heavier iron particles.

Where the iron is in the form of nodules, much effort is put into separating the iron-bearing part of the ore from the unwanted gangue (see glossary). The ore is first roasted; this serves many purposes, amongst them to drive off any water, carbon dioxide or sulphur present in the ore and to make the ore more porous. This makes the ore more reducible and to enable hard ores to be broken up. Roasting is usually carried out in open fires or shallow pits, but never in kilns. Water present will result in explosions in the fire, and if the roasting was in a confined space such as a kiln the effects would be disastrous. The charge has to be of a size that will allow the reducing gases to penetrate the ore and bring about the reduction of iron. Fine iron ore is added to the furnace in small amounts at regular intervals. This is the case with other ores too, where the reason is to allow the draft to penetrate. The ores are very rich in iron as can be seen from analysis of the resulting sands, that which contain 60-70% iron oxides.

Furnace types

The division of the different types of furnace are dealt with in section 2.3 of this book, but for the purposes of this section we can divide furnace types into bowl or shaft, and forced draft and natural draft. Tylecote (1965, 341) has asserted that few pre-industrial furnaces are of the bowl type, where the bloom is withdrawn from the top of the furnace, and that the vast majority are of the shaft type, where the bloom is removed from the front of the furnace. Tall shaft furnaces have natural draft introduced by the chimney effect which causes air to be drawn into the furnace. Low-shaft furnaces require the air to be forced in with the aid of a blowing device, usually a pair of bellows.

The use of fluxes in smelting

In modern iron-making, fluxes such as limestone are added to the charge to convert unwanted parts of the ore into a liquid that can be tapped off from the furnace. Todd and Charles (1978, 64) report that some slag from the discard of previous reductions, and selected by its higher density, was added to the ore. This may have been to act as a fluxing agent, but in general fluxes are not added in primitive iron smelting.

Yield

In order for iron ore to be worth smelting by pre-industrial methods the percentage of iron that can be extracted must be in the region of 20% and upwards Coghlan (1977, 9). Tylecote (1962a,182) has suggested that the best ores for smelting are not necessarily those with the highest iron content but those low in silica, alumina, lime and magnesia.

Forging

The product of most primitive furnaces is called a bloom. This is a heterogeneous mass of slag with reduced lumps of porous iron, pieces of charcoal and voids. To produce workable iron from this mass, the slag and other impurities have to be expelled and the voids closed. The bloom is heated to a temperature at which the slag is fluid and the disperant pieces of iron will weld together under the force from hammer blows. There are various methods of doing this. The bloom can be forged as a single mass or it can be cut into small pieces and then forged into workable iron by heating and hammering. Alternatively the bloom can be broken into small pieces and the iron and slag separated by hand. These pieces are then welded together and forged into workable iron. The broken pieces are wrapped in a clay envelope and heated to welding temperature (the clay prevents oxidisation) and are then welded to form a billet.

Modern technology

Modern technology can be used to gain insight into ancient practices. Modern methods are a matter of fact, and, to some extent, so is the recent history of modern iron-making. A study of the metallurgy and chemistry of modern methods will reveal in minute detail exactly what takes place. This knowledge will allow speculation as to what other methods could have been used to produce a particular type of ferrous metal.

Observation of the practical resources and operational methods of modern iron making will reveal the working practices. With most modern iron production, much of the work is done by machinery. It is not safe to simply assume that when considering technology in antiquity manpower can be substituted to take the place of machinery.

An example of this can be found when considering the refining of bloom iron. In the modern practice the bloom iron is placed under a power hammer to weld it together. It is easy to see that an observer may well propose that before power hammers, manpower could be used to achieve the same results. This is not so. In the case of bloom iron, the surface area of the hammer has to be equal or larger than the size of the bloom in order to weld it together. This is not possible with hand-held hammers.

The most common view is that in antiquity, all the power for conducting manufacturing processes could be supplied by manpower alone. It is undoubtedly true that at various periods in the history of the Roman empire there were large amounts of man power available, but not all the time, owing to the fortunes of war. It will be shown in chapter 5 that no amount of manpower would be able to produce iron wire in the quantities and to the accuracy that is shown in Roman mail coats. The problem is solved by the application of technology not that of man power. Further, there are occurrences of what can only be described as factories such as those at Barbégal in France (Hodge 1990). It is not a justification to say that instances of water power are from a late period and that before this time manpower was used. The instances found may be from a late period but it is equally possible that water power was used before and its use has yet to be identified.

It is essential to use whatever evidence is available as a guide to helping unravel the problems of ancient technology. It is not justified to simply substitute manpower for machine power and think the question ends there.

1.6 Summary

This chapter has shown how fundamental to the Roman world iron was. For all its importance, very little is known about the way in which iron ore was mined and processed to produce the wrought iron and steel artifacts which are found in the archaeological record. This book describes in detail experiments which were conducted to replicate Roman blacksmithing practices and manufacturing techniques. Chapters are devoted to considering the technical aspects of iron making: bloom forging; the production of a range of artifacts; and an explanation of the by-products from production.

CHAPTER 2

The technical aspects of iron making and a review of experimental iron making

'It seems to me an art comprising great knowledge, for I know of no art or activity whatever, excluding the sciences and painting, that does not need this as its principal member. Therefore in my opinion, if it were not for the nobility of the material, I would say that the smith in iron should justly take precedence over the goldsmith because of the greater benefit he brings.'

Biringuccio, Pirotechnica, 1959, 373-374

2.1 Current approach to experimental iron making

Ancient iron making differs in many ways from its modern counterpart therefore, in order to avoid confusion between ancient and modern methods, a brief outline of the two basic processes follows. They are the *direct method*, or bloomery process, and the *indirect method*. In the direct method the iron is made in one operation, and is produced as a solid mass of sponge-iron which is then forged into workable iron. The direct process is expressed by the formula $Fe_2O_3 + 3CO \rightarrow 2Fe + 3CO_2$. In Europe, this method was used from the time iron was first smelted until the 15th century. The relatively modern indirect method is a high-temperature process in which liquid iron is produced and is then further processed into other ferrous products such as cast iron and steel.

Iron was first smelted by the direct process. The chemistry involved has been explained by modern scientists, but much of the technology and skills necessary to produce iron have been lost. There are no written records available from the early periods (Cleere 1971, 203) and so we must look for other sources of information.

A study of pre-industrial methods of iron production will give indications of the technology available to the Roman blacksmith. As mentioned in chapter 1, two methods which give a useful insight are the Japanese Tatara method and the African Maffa method.

Japanese Tatara Method

This process is still in use today and seems to have come to Japan from Manchuria by way of the Korean peninsula in the sixth or seventh century (Kapp et al. 1987, 62). The process would fit into the class of slag tapping, induced draft type of furnace but, because the furnace itself is a rectangular clay box (547 cm long, 152 cm wide and 121 cm tall with 25.4 cm thick walls), it is difficult to fit it into any of the categories of furnace mentioned under Furnace Types in this chapter.

The iron ore is supplied in the form of a black sand called *satetsu*, which contains only about 1% of iron. The iron present in the sand is separated out from the other silts by washing. Air is introduced by a series of tuyères near the base of the furnace.

The whole cycle of smelting takes five days - one day to build the walls of the furnace from clay and sand, three days to smelt, and one day to remove the iron. A charcoal fire is started in the furnace and when the charcoal is at the correct temperature a layer of iron sand is spread over the coals, followed by a layer of charcoal. After thirty minutes another layer of sand and charcoal is added. This sequence is continued for the next seventy-two hours. Rostoker et al. (1989) states that an eighty-four hour smelt consumed 22 tons of charcoal and 20 tons of *satetsu*, giving a 50% recovery of iron from ore. The finished product is called the *kera* and is composed of approximately 50% steel ranging from 0.6%C to 1.5%C.

Mafa down-draft furnace

The second case of a whole range of ferrous metals being produced in a single operation is in the Mafa down-draft furnace (David et al. 1989). The furnace is a shaft 30 cm across, 30 cm deep, and 162 cm high, built of a base of dry-stone walling, a mid-section of alternating courses of stone and daub, and an upper section of coils of daub. The inner surface is plastered with daub. Air is supplied from a single tuyère that is introduced through the top of the furnace and hangs vertically down in the centre of the furnace. The tip of the clay tuyère is intended to melt and its molten by-products play a vital role in ore reduction.

The actual recovery rate of iron from ore was 61%. This is attributed to the high grade of iron ore used. Half of the iron pellets recovered from the smelt were found to be cast iron, whilst the remainder were steel and low carbon iron. David et al. (1989, 198) state that the cast iron was a deliberate product because of the methods that were used to decarburize and make it suitable for forging.

The Mafa and the Tatara methods, although from different continents, share the common feature that both produce the whole range from low carbon iron to cast iron in one operation; the direct method.

2.2 Discussion

The assumptions of earlier writers on iron production such as Singer et al. (1954, 56), that furnaces were not capable of attaining high enough temperatures to produce cast iron have been largely disproved by Tylecote et al. (1971) and Rehder (1986). Their work has shown that a simple shaft furnace is capable of reaching temperatures of 1600°C and at this temperature cast iron can be produced. It has also been shown that a working temperature of 1200°C will produce bloom iron. Though only slight attention has been given to the further processing of bloom iron, the work that has been done shows this to be a lengthy process involving many man-hours and a large to be a lengthy process involving many man-hours and a large consumption of fuel.

Another source of information is to conduct experiments with reproductions of early iron-making equipment. No one set of experiments will comprehensively answer all or even most of the problems of early iron-making. The discussion above demonstrates that furnace types and materials vary a great deal from country to country, and even region to region, as do the kinds of fuel and iron ore used. I shall give a short summary of the work carried out generally, but concentrate on the experiments conducted in Britain.

The following survey will demonstrate the current level of knowledge concerning the smelting of iron and give an indication of the nature of the bloom iron produced in antiquity.

Bloom iron is the starting point for the experiments described in the following chapters. The working parameters set for those experiments were arrived at on the basis of the knowledge gained from the various smelting experiments. Experimental smelting has been carried out by a number of researchers. Tylecote, Austin and Wraith (1971, 363) discuss work by themselves and Gillies (1957); Wynne and Tylecote (1958); Pliner and Radwan (1962); Radwan and Pliner (1963); Thomsen (1963); Straube (1964); Radwan (1965); Hagfeldt (1966); Pliner (1969); and Cleere (1971). Other work has been performed by Souchopova and Stransky (1989) Pliner (1991); and Crew (1991).

Wynne and Tylecote (1958)

Wynne and Tylecote (1958) conducted a series of experiments in a simple bowl furnace using tuyères made from silica, and air supplied from an electric compressor. The bloom produced was described as unworkable and was subsequently heated in a crucible inside a gas-fired furnace to melt and drain the slag out. There is a brief description of the forging but no indication of the time involved. The yield was given as 0.34 kg of iron from 1.81 kg of ore with a charcoal consumption rate of 7.26 kg per 0.45 kg of iron. This was a non-slag-tapping furnace (Fig. 2.1 Group A).

Cleere (1971)

Cleere (1971) describes experiments conducted to smelt iron by the direct process in a reconstructed Roman shaft furnace. He considered the determining factor in the process was not the melting point of the metal, or the reduction temperature, but the temperature at which fluid slag could be obtained. His experiments showed that the most efficient mode of operation for this type of shaft furnace is for the slag to be removed continuously from the hearth without heat loss. Valuable information was also discovered about the manning requirements needed to operate this type of furnace. The minimum manning requirement per basic unit of ore roaster and smelting furnace would have been ten men. In the most successful of the trials 9.07 kg of bloom iron were produced from 91.17 kg of siderite (mudstone) ore and 116.35 kg of charcoal in 10 hours. The bloom was smithed, but few details are given and there is no information about the time involved. As the main objective of the experiment was to produce bloom iron, the subsequent processing (bloomsmithing) was not reported in any depth, other than a description of the condition of the finished metal.

Tylecote, Austin and Wraith (1971)

Tylecote, Austin and Wraith (1971) reported a series of 32 experiments. Iron smelting was conducted in a copy of a second century slag-tapping shaft furnace found at Ashwicken, Norfolk (Fig. 2.2). The purpose of the experiments was to investigate the variables affecting the bloomery performance in order to derive an effective smelting routine. The experiments showed that 8 kg of iron could be produced from 50 kg of ore and required 40 kg of charcoal.

Crew (1991)

In 1985 Dr Peter Crew started a series of experiments whose main objective was to provide comparative data for the interpretation and quantification of the iron-working debris recovered from two late prehistoric sites in Merioneth. The experiments were conducted using reconstruction's of non slag-tapping shaft furnaces. The initial experiments were unsuccessful but later experiments produced bloom iron that was smithed into bar iron. Accurate descriptions of both bloomsmithing and blacksmithing were given. Some of the later work on the bloom was conducted by a professional blacksmith who used a modern forge hearth, mineral coal as fuel, and a power hammer to complete the forging of the iron. Using his own methods of experimentation, Crew estimated that one kilogram of fully smithed bar iron required 100 kg of charcoal and 25 man-days work (Crew 1991, 35). Crew (1993, 2) gives a brief description of an experiment using a slag-tapping shaft furnace similar to the shaft furnace shown in Fig 2.2. A 1.8 kg bloom was produced and forged into a compact block in 1.5 hours. At the time of writing there are no further details. The objectives of the various experiments vary but few seem to concentrate on the further processing of the bloom.

As the above review demonstrates previous experiments with the exception of Crew's work, have paid only slight attention to the process of bloomsmithing. The product of any furnace operating on the principle of the direct reduction of iron is a bloom of iron. The bloom is totally unworkable. It requires considerable skill to turn bloom iron into workable iron. As will be shown in chapter 4, iron was probably delivered to the blacksmith in the form of billets (see glossary). These had to be turned into bars which were then forged into artifacts. The time needed to accomplish this, and the associated technical difficulties, have not been addressed in the existing literature. The transformation of iron ore into finished artifacts involves a series of different operations. The smelting process is only one of them, and should not be seen as an end in itself, but as only one of the many stages which must be completed to produce iron artifacts. It is one of the primary objectives of this book to consider the technology associated with the Roman-British iron industry. A principal characteristic of Roman techniques is the tapping of slag from the furnace as a fluid, which helps distinguish Roman technology from earlier processes (Tylecote 1976). Using this criterion the

Austin and Wraith (1971), are more likely to yield the type of information which will be close to the Roman method of iron production. A clear understanding of the ancient methods of iron making is essential, because the form of the iron after smelting has a direct influence on the further processing necessary to refine it. Consideration of this refining process is a major component of this book. The methods of iron smelting considered in this section are restricted to the technology that was thought to be available in Britain during the Roman occupation i.e. the direct process (Tylecote 1976, 53-57).

The process of iron-making is thought to be as follows:

1. Mining to produce iron ore.
2. Iron ore is roasted, using wood as a fuel (only certain ores need this treatment).
3. Wood is burnt to produce charcoal.
4. Iron ore and charcoal are heated in a shaft or developed bowl furnace and a bloom of iron is produced, with a quantity of slag.
5. The bloom is heated and hammered to expel the slag and weld the iron particles together. This produces a billet.
6. The billet is heated and hammered to produce a bar.
7. The bar is heated and hammered to produce artifacts.

2.2 Detailed description of the direct process of iron making

After mining the ore is roasted if required. Ores which require roasting will either be carbonate ores, in which case roasting drives off CO_2 or limonite ores which contain $Fe_2O_3H_2O$ and water which needs removing. Roasting makes the ores more porous (and hence more easily reduced) and allows some hard ores to be broken into small pieces (again making them more reducible) (Healy 1978, 183).

The reduction process lasts about 15 hours. The iron ore is stacked in layers with wood in a pit which could be one of two types, either bowl or trench. The ore is roasted to a temperature of 500°C. The pit is needed because during heating the ore can explode with considerable and spectacular violence due the presence of water. Such an explosion in a smelting furnace would be disastrous. I assisted in an experiment to roast bog iron (Snowdonia National Park study centre 1991). After roasting the ore could be broken with bare hands. When the initial fire burns down, the ore is left until the fire goes out. Some skill was needed in managing the fire to ensure uniformity in the roasted ore.

Direct reduction

The direct reduction of iron from iron ore depends on the production of carbon monoxide from the fuel which then combines with the oxygen in iron oxide; producing the gas carbon dioxide and leaving behind pure iron. The carbon dioxide will escape from the top of the furnace. Raw iron ore is not pure iron oxide, but is made up of iron oxide and other unwanted minerals, referred to as gangue. The gangue has to be removed by slagging in the smelting process. Iron oxide will reduce to iron at a temperature of 800°C (Tylecote 1962 b, 183-4), much lower than the melting point of iron which is 1540°C (Kempster 1964, 6). The most significant factor in the process is the temperature at which the slag becomes fluid, 1150°C (Cleere 1971, 204). In practice the operating temperature of a bloomery furnace is in the region of 1200°C. At this temperature, the furnace is hot enough for the slag to become a liquid and drain away from the iron, leaving behind a spongy mass known as a bloom.

Furnace types

The exact nature of the various types of furnace used in antiquity is still unclear because few furnaces are found in good condition and the upper parts are usually missing, therefore the shape of the furnace above ground level is largely a matter of conjecture. Although distinction between furnace types is not exact, the following types are well established. Coghlan (1977) distinguishes between the simple bowl, the domed or pot furnace and the shaft furnace (see Fig. 2.1). Cleere (1972) classified furnaces on the presence or absence of slag tapping. His types are divided into groups:

1. simple bowl;
2. domed or pot furnace: and,
3. shaft furnace.

Group A non slag-tapping furnaces

Diagnostic features:

1. No provision for tapping of molten slag.
2. Hearth below surface of surrounding ground.
3. Blown by forced draught.
   Sub Group 1. No superstructure (bowl furnace).
   Sub Group 2. Superstructure - cylinder or truncated cone.

Group B slag-tapping furnaces

Diagnostic features:

1. Provision for tapping of molten slag.
2. Hearth level with surface of surrounding ground.
3. Superstructure
   Sub Group 1.i.
   Blown with forced draught. Cylindrical superstructure.
   Sub Group 1.ii.
   Blown with forced draught. Conical or hemispherical superstructure.
   Sub Group 2.i.
   Blown with natural draught. Cylindrical superstructure.
   Sub Group 2.ii.
   Blown with natural draught. Conical or hemispherical superstructure.

The low shaft furnace of the type found in Eastern England (Fig 2.2) was the principal type used in the Roman period Tylecote (1987, 160).

Figure 2.1 Sub Group 1.i. illustrates Cleere's group B slag-tapping furnace. The bowl furnace (group B non slag-tapping furnace Sub Group 1) was in constant use in the whole of Britain throughout the Roman era (Schubert 1957, 44). It seems possible that non-slag tapping furnaces were in use in areas outside the

Roman sphere of influence, or where iron production was for local use. Tylecote (1976, 54) states that the bowl furnace developed in height and diameter until it had more than one tuyère (see glossary). Figure 2.3 shows a developed bowl furnace (Cleere, Sub Group 1.ii) of the Roman period.

As Fig. 2.3 shows, it would be an easy task to extract a hot bloom from a furnace with such a large opening at the top. A large bloom could be produced from this type of furnace, and a large output seems likely. No references have been found to experimental smelting being carried out in this type of furnace, so its operating parameters are uncertain. The size of the iron industry in Britain, and with it the production of iron increased on an enormous scale with the Roman invasion. In pre-Roman sites the slag is measured in hundreds of kilograms, while slag from sites of the Roman period is measured in hundreds of tonnes. It seems likely that this increase was due to improved production techniques, such as slag tapping furnaces (Tylecote 1976, 53).

The significance of furnace material

All of the materials present when smelting in a primitive iron smelting furnace - ore, fuel, furnace materials - will contribute to the physical nature and chemical composition of the bloom. As the intentional addition of fluxes in smelting is not attested until the Middle Ages (Cleere 1971, 205), the materials present in Roman smelting are iron ore, fuel and the furnace lining.

Fulford and Allen (1992, 197) have shown that furnace linings were significantly consumed in the primitive iron-making process. This was certainly true when rich ores were used. "The interior of furnaces were relined with clay of 10 mm to 30 mm thickness before refrying. It was added to the surface of glassy, vesicular clay created during the previous smelting. The clay was altered mineralogically and changed physically to a glassy liquid which sloughed off the furnace wall and so could enter into and react with the charge of iron ore and charcoal. Analysis of the non-ferruginous matter in the slags shows that it was derived from the furnace lining". They add that if clay had not been derived from the furnace walls, it would have been necessary to add it. The lining of the furnace clearly plays a vital and active role in the slagging of the ore.

Since ores, fuel and the clay used to make furnace linings differ from one site to another, there will be small differences in the chemical composition of tap slags from different sites. In any analysis of residues from iron-making sites, the nature of the furnace lining plays a vital role.

Fuel

The most important single fuel and reducing agent in the ancient world was charcoal. Charcoal is the residue obtained as a result of the incomplete combustion of either vegetable or animal materials. At the end of the process the material is almost pure carbon. The volume of the original wood shrinks to two thirds and the weight is approximately 25% of the original. Charcoal is capable of burning at a temperature of 900°C but the use of an air blast will enable temperatures of 1600°C to be reached readily in a furnace (Rehder 1986, 91). In Britain oak, birch, ash, hazel and maple were all used to produce charcoal (Healy 1978, 150). For the smelting of iron, charcoal was the principal fuel. Coal is not suitable for smelting iron because the sulphur content makes the iron too brittle for forging. It is known that coal was used for some industrial processes and its use is recorded on 51 Roman sites (Webster 1955). Coal can certainly be used for the forging of iron, indeed it is a superior fuel and there is no doubt the Roman smiths made use of it. In most places where there were mineral coal outcrops it was exploited by the Romans (Schubert 1957, 44).

Iron ore

In Britain, iron ores are widespread and of many kinds, and have been mined in many places at one time or another. The main types of ore found in Britain are carbonate (siderite), oxide (e.g. haematite) and oxy-hydroxide (e.g. limonite). The carbonate ores are the most common and occur as nodules (clay ironstone) in the Wealden Series or as sedimentary deposits, siderite or spathic iron ore in Northamptonshire, Lincolnshire, Oxfordshire and the Cleveland Hills. Haematite occurs in West Cumberland and the Furness district of Lancashire. Limonite is found in the Forest of Dean and the edges of the South Wales coalfields.

Most deposits of non-ferrous ores yield some iron ore of various types. Ores at Alderley Edge occur with copper, at Weardale and the Mendips with lead, and in the Leadhills of Lanarkshire with numerous minerals.

Figure 2.4 shows the general distribution of the location of ore roasting pits and where mines have been found. Figure 2.5 gives a general distribution of sites where smelting furnaces, smithing hearths and unidentified features have been found. A gazetteer of Roman-British ironworking sites has been compiled by Aiano (1977). In the Forest of Dean, an important centre of iron production was Ariconium (modern Weston-under-Penyard) (Jones and Mattingly 1990, 193). In the Roman world as a whole, iron ore was mined in Spain, Gaul, Italy, Elba, Sardinia, Sicily, Central Europe, Noricum, Illyria, Macedonia, Asia Minor and Africa (Healy 1978, 64-65).

Temperatures in a primitive iron smelting furnace

Although it is possible to reach temperatures of 1600°C in a shaft furnace a temperature of 1200°C is all that is required to bring about the production of iron by the direct method of iron making. Tylecote et al., (1971, 363) make a review (their table XIX) of experimental smelting. The experimenters summarised here are listed with the temperatures attained: Gillies (1957) 1300°C; Wynne and Tylecote (1958) 1250°C; Pliner and Radwan (1962) 1300°C; Radwan and Pliner (1963) 1300°C; Thomsen (1963) 1270°C; Straube (1964) 1420°C; Radwan (1965) 1300°C; Hagfeldt (1966) 1300°C; Pliner (1969 and 1991) 1545°C; Cleere (1971) 1300°C; Tylecote, Austin, and Wraith (1971) 1400°C.and Souchopova and Stransky (1989).

Pliner (1969) achieved a temperature of 1545°C, at which temperature cast iron can be produced. Rehder (1986) argues that a simple charcoal furnace 30 cm in diameter with a single tuyère easily powered by one man can reach a temperature of 1600°C. Tylecote et al. (1971) stated that a shaft furnace is capable of producing metal ranging from wrought iron up to cast iron, and that by altering the fuel-ore ratio, steel can be produced. A simple furnace, if handled correctly, is capable of producing the whole range of ferrous metals. Early furnaces were capable of producing cast iron, but there is little evidence to support the use of cast iron in the Roman period. The capability to produce steel also existed in Roman times, but it seems likely that good quality steel was imported (see chapter 1.4).

The shaft furnace is thought to be a common type of furnace in use during the Roman occupation of Britain. Shaft furnaces were often constructed into banks. They were in the form of hollow cylinders, with an arch opening at the bottom. The arch was used for both the drainage of slag and for pumping air to heat the fire, possibly using bellows. It is possible that the bloom was extracted through the arch.

The furnace is charged with alternate layers of iron ore and charcoal. The charcoal burns and gives off carbon monoxide gas which reacts with the oxygen in the iron oxide and forms carbon dioxide, leaving behind iron. As the temperature in the furnace rises, the materials in the ore melt. The impurities form a liquid at about 1135°C, but iron does not become a liquid until 1530°C, so at a temperature of 1200°C the iron is not a liquid, but forms droplets which concentrate near the tuyère. The liquid slag collects at the bottom of the hearth and is tapped off, in either a continuous or intermittent flow. This process is continued until the bloom which builds up below the tuyère is so large the air will not circulate and the process stops.

2.3 The composition of modern wrought iron

As has already been stated, the most common ferrous metal in Roman times was the metal that we call wrought iron today. However, although the methods of producing the metal are similar, they are not identical. Modern wrought iron is a matrix of pure iron with slag dispersed through the structure. It is produced by a process called puddling. In this process the charge (cast iron) is placed in a furnace with a domed roof which reflects heat down onto the charge. When the charge is melting, it is stirred by hand with long iron rods, made from heavy bar iron. The author assisted in a smelt and had the opportunity to try puddling. The heat from the furnace saps the energy at such a rate even the experienced puddlers can only work for three to four minutes at a time. The quality of the finished wrought iron depends at this stage on the experience of the puddlers, in knowing exactly how long to stir the iron before it is removed from the puddling furnace. The mass of iron is placed under a power hammer, and is 'shingled'. This involves squeezing the mass of iron to fuse it and expel slag. At this stage, slag can be seen running out of the iron in large quantities. The iron is squeezed on both sides. A power hammer is then used at its full force, expelling slag at great speed and in large quantities, giving the impression that it is raining liquid metal. When the bloom has been forged into a rectangular block, it is placed between sets of rollers and rolled into flat strip. The material produced at this first rolling is called muck bar. It must be kept in mind that the hammering and rolling are done without reheating the iron, so speed is of the essence. The force applied to the mass of iron causes the slag inclusions to decrease in cross section and increase in length to form long streamers which are parallel to the direction of working. The rolled strips are then cut into short lengths and put into piles. These piles are heated in a balling furnace until the heat causes the metal to fuse. This pile is hammered under a power hammer, rolled into strip again and cut into short lengths. These are re-heated in a balling furnace and the process is repeated. The quality of the iron is determined by the number of times the process is repeated. Most of the refining has to be carried out at or near welding heat (1150°C to 1200°C) and at these temperatures large amounts of iron are lost due to oxygenation of the surface layers.

Being able to work at first hand on such a process gave a valuable insight into the energy required by the workforce who made their livings in this manner. It was an exhilarating experience, and showed the dangers and sheer hard work of this process. At the end of the day even the experienced workers were grateful for a rest. It also provided a valuable insight into the nature of the metal and gave a greater understanding of why wrought iron behaves the way it does when it is being forged.

In the Roman period it has been supposed that the process of converting bloom iron into wrought iron was by heating and hammering in the same manner as described above but with the substitution man power for machine power. The experiments described in chapter 3 show that this is not as straightforward as it at first appears. The manpower and time involved in bloomsmithing is substantial and simple substitution of machine by manpower alone does not overcome such problems as the bloom breaking up during forging or the time taken to heat a large bloom to a high enough temperature for forging.

2.4 Refining the bloom

The product of the bloomery process is called a bloom, which is a mass of pure iron interspersed with slag. In this condition the iron is unworkable and has to be refined to produce iron which can be forged into artifacts. The objective of the next stage in the cycle is the removal of slag and the welding of the iron into a solid mass.

For welding to take place, the surfaces to be joined must be clean and have as little oxide on them as possible. Iron oxide has a higher melting point than iron and acts as a barrier between the pieces of iron which are to be joined by welding. The presence of slag between the particles of iron also prevents the iron from welding. However, the slag binds the bloom together and keeps it as a solid mass, so it has to be removed in such a way that it is expelled without the bloom collapsing. This process is called bloomsmithing. It is carried out by heating and hammering the bloom to expel the slag and consolidate the iron by welding the particles together.

The forging temperature must be sufficient to cause the slag to become fluid enough to be expelled during hammering, but not so liquefied that it flows out of the bloom on its own. If this happens it creates voids between the particles of iron, and the first blow with the hammer will cause the bloom to fall apart. If the traditional method of bloomsmithing is followed, then great care is needed in the initial stages, and heavy blows will cause the bloom to fall apart. The process will continue until a consolidated bloom is produced. Experiments (see chapter 3) have shown this to be a very highly skilled and time-consuming process.

2.5 Forge welding

To weld is, by the dictionary definition, to "unite (pieces of metal, esp. iron) into homogeneous mass by hammering or pressure (usu. when iron is softened by heat but not melted)". (Concise Oxford Dictionary 1985, 1221). Welding in antiquity was by the process that has become known as fire welding (see glossary), or forge welding. The pieces to be welded are laid on top of each other and a heavy blow is struck with a hammer. It is vital when conducting fire welding to have the iron at the right temperature. For wrought iron, the metal is heated to a temperature of 1100°C, at which temperature the metal is very soft but not molten. It is in its most plastic condition (see glossary) and this ensures it is soft enough to weld together when subjected to force. The blow causes the metal parts to weld together and become a solid mass. Every time a fire weld takes place, large amounts of metal are lost because of oxidisation - often as much as 60%. To ensure enough metal remains, the welded parts are made larger than the finished size. The surfaces to be joined are often coated with a flux (see glossary) to help dissolve the oxide film and prevent further oxide from forming.

It has not been possible to prove the use of fluxing in ancient forge welding. A series of three experiments was carried out by the author to determine if forge welds carried out using different fluxes left any identifiable residues in the welds. Scanning Electron Microscope (SEM) analysis of all welds showed no traces of fluxes left in the welds. However, it is known that in the Roman period, fluxes were used for both hard and soft soldering of bronze, brass and iron (Aitchison 1960, 214); therefore as the necessity for fluxing was recognised, it is possible that fluxing was carried out in the fire-welding of iron.

The successful makers and shapers of iron learned to sense many factors that we need laboratory methods to detect (Gordon 1991, 135). This is conspicuously so in fire-welding, because it relies on the skill of the operator to be able to judge by eye the exact moment that the metal is at the correct temperature for welding.

Fire-welding is, and always has been, a highly skilled operation requiring considerable practice to reach a level of proficiency. If the temperature is too low, the parts will not weld; if it is too high then the metal will burn and the resulting weld will be brittle. If the weld is carried out correctly, then the welded section will be as strong as the parent bar. The interface of welded sections shows as a line on the surface and as a white line on X-rays. If a bar made from welded pieces is given enough forging then the weld lines will become incorporated in the matrix of the metal and will eventually disappear completely, making it impossible to tell if the bar has been made from one piece or several.

2.6 Discussion

The assumptions of earlier writers on iron production such as Singer et al.(1954, 56), that furnaces were not capable of attaining high enough temperatures to produce cast iron have been largely disproved by Tylecote et al. (1971) and Rehder (1986). Their work has shown that a simple shaft furnace is capable of reaching temperatures of 1600°C and at this temperature cast iron can be produced. It has also been shown that a working temperature of 1200°C will produce bloom iron. Though only slight attention has been given to the further processing of bloom iron, the work that has been done shows this to be a lengthy process involving many man-hours and a large consumption of fuel.

In general, and in what we have discussed so far, it has been assumed that the processing of iron involves first smelting iron ore to produce a bloom, which is then refined by heating and hammering, but the technical details remain obscure. Rehder (1986) has suggested that a two-step cast iron finery procedure (see glossary) was possible from very early times, and it seems likely to have been used more widely and earlier than currently thought. He also suggests that the melting of metals can be performed in mixed fuel furnaces which are identical to smelting furnaces, but at rates 6 to 10 times faster and consuming proportionately less fuel. He goes on to say that The final iron products of the bloomery and the finery processes are identical and there is no clear way of telling by which process an iron artifact was made (Rehder 1986, 91).

Until now researchers have assumed that all iron produced in the Roman period was bloom iron, but there is a possibility that a finery process (see glossary) could have been used. If it were, then the amount of iron required could have been produced in considerably less time and at a much lower cost than has hitherto been considered. The Romans did not consider iron to be an expensive metal (Schubert 1957, 34). Large amounts were abandoned, such as the deposits found at Inchtuthill. The number of metal objects found must surely seriously under-represent the quantity in circulation (Rehder 1986, 91). If iron was indeed a cheap metal, and large amounts were in circulation, it is possible that a more efficient production system than the bloomery process may have been in practice. The size of bloom produced in the Roman period has a significant effect on the equipment needed to process it into workable iron. Tylecote (1976, 55) has stated that the size of bloom in the Roman period varied from 5 kg to 10 kg. Singer et al.(1954, 56) asserts that the Roman furnaces of Noricum rarely produced bloom of more than 45.36 kg. From this it may be assumed that blooms approaching 45 kg were made. A large bloom would present a considerable challenge in heating and handling if it was forged as a whole. Bloom iron is difficult to heat because the slag inclusions make it a poor conductor of heat. A cold bloom of 50 kg will require a large furnace and considerable amounts of fuel to reheat it (see chapter 3). Even a 10 kg bloom will require considerable heating and is very difficult to manipulate. Both problems are overcome if the bloom is broken into small pieces for further processing.

The main obstacle which slows down the iron-making process is the removal of slag from the bloom. Crew (1993) has shown that bloom from a slag-tapping furnace, can be processed in a shorter time than that from a non slag-tapping furnace, although this is only borne out in one experiment. The introduction of another heating process after smelting which would remove large amounts of slag before bloomsmithing took place, might well produce an iron that can be more readily processed into workable iron.

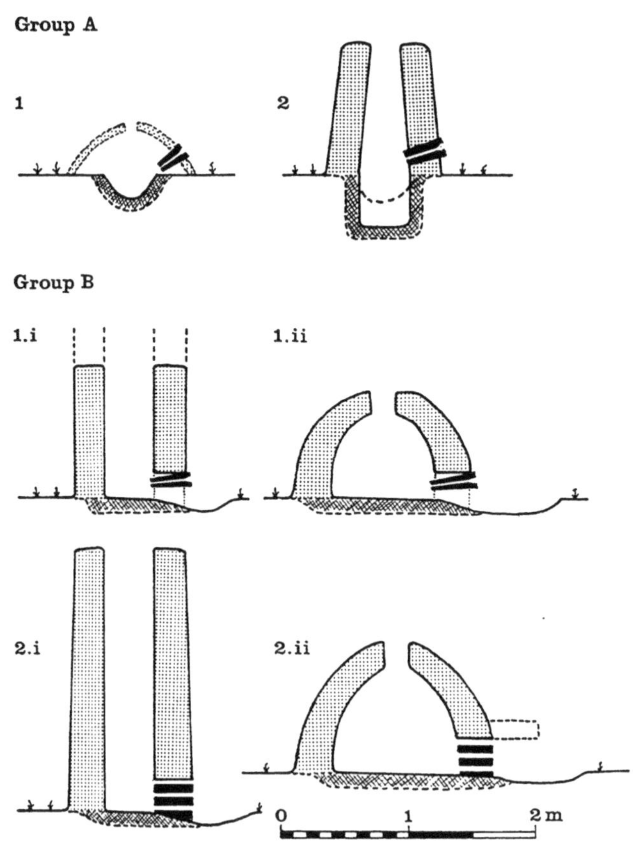

Figure 2.1 Sections through early iron-smelting furnaces. (Cleere 1972. Fig. 11)

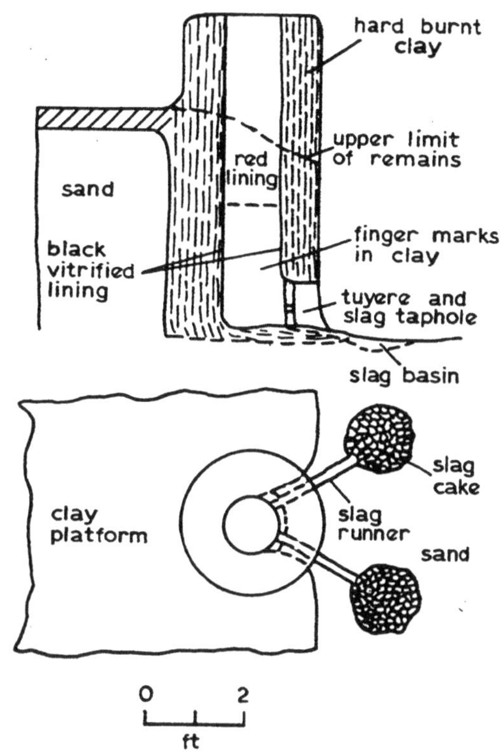

Figure 2.2 Plane shaft furnace of the Roman period. (Tylecote, 1962b Fig. 51)

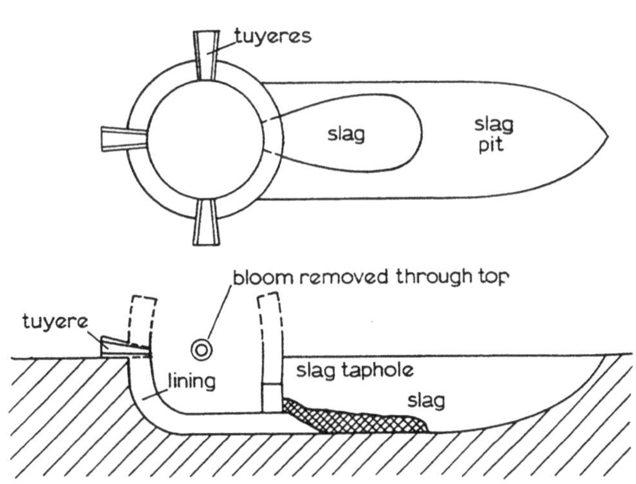

Figure 2.3 A developed bowl furnace of the Roman period - Sub Group 1.ii. (Tylecote, 1976 Fig. 41)

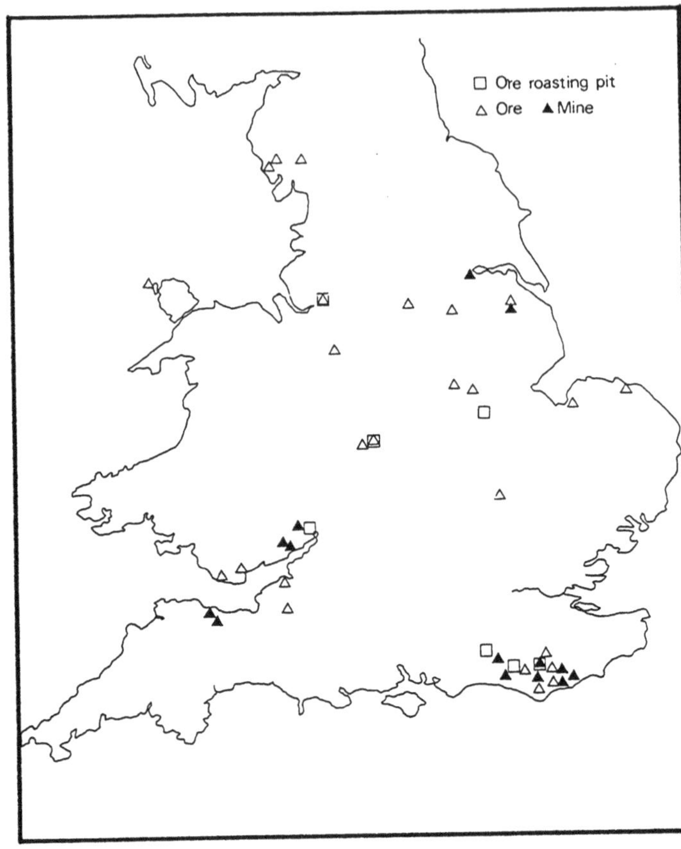

Figure 2.4 General distribution of the sites where iron mines, ore and ore roasting pits have been found. (Aiano, 1977 Fig. 74)

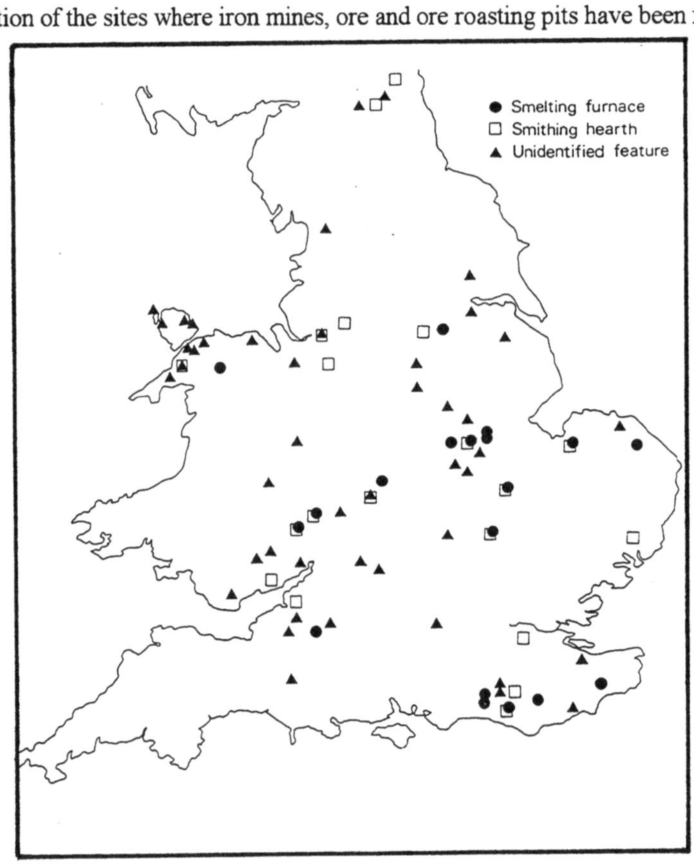

Figure 2.5 General distribution of sites where smelting furnaces, smithing hearths and unidentified features. have been found. (Aiano, 1977 Fig. 76)

CHAPTER 3

Bloom forging experiments

'Iron, cold iron, is master of men all.'

Rudyard Kipling, *Cold Iron*

3.1 The Refining Process

In Chapter 1.1 the objectives of this set of experiments were stated as follows:

(a) To determine the amount of time needed to process bloom iron into workable iron, that is, iron which can be forged into artifacts.
(b) To arrive at a design of hearth suitable for smithing blooms that is comparable with the archaeological record.
(c) To quantify the amount of charcoal used.

During the course of these experiments hammer scale was collected for analysis. To compile distribution maps, residues from the forge hearths were also collected for analysis. The results compiled from this data are presented in chapter 5.

I have already discussed previous experimental bloom-iron production using reconstructed furnaces of various designs cf. Wynne and Tylecote (1958), Cleere (1971), Crew (1991) and Senn and Fasnacht (1991). Their main objectives were to investigate the workings of furnaces used to smelt iron, and to produce bloom iron. The further processing of the bloom was a secondary consideration and refining or bloomsmithing was conducted either by the experimenters themselves, or by a professional blacksmith.

Cleere (1971) gives very little information on bloomsmithing. He notes that the bloom was worked by repeated heating and hammering into small blanks. Wynne and Tylecote (1958) used a professional blacksmith to process the iron they produced, but give few details of the forging of the bloom.

Most modern blacksmiths have been taught to work on mild steel and have little or no experience of wrought iron, much less bloom iron. This places them at a considerable disadvantage. Their comments, although valuable, are based on working a material of which they probably have had no experience.

When a researcher from a discipline such as archaeology attempts bloomsmithing, unless he is a skilled blacksmith, it will only be possible to copy what has been seen on film, or what has been learned from literature. The many subtleties of iron working will be unappreciated. For example, a considerable amount of practice is necessary before a blacksmith develops the ability to feel the movement of the iron through the hammer, and not rely on eyesight alone. Feel is important because visually the iron may not appear to be very different, but the ability to feel, for example, developing internal cracks can be of considerable benefit. It also takes time to learn the subtle changes of colour that take place as the iron is cooling. These are another indicator of when the iron has become too cold to work. To an unskilled eye there will be no change in colour, and the novice will go on working after the iron is too cold. If this is done when bloomsmithing, the bloom will fall apart. The experienced ear can detect a range of sounds and this information is vital in order to understand the state of the iron being worked. Although Crew (1991) conducted the early stages of bloomsmithing himself, the later stages were carried out by a professional blacksmith using modern fuel and tools. Senn and Fasnacht (1991) also used a professional blacksmith who likewise used modern fuel and tools. The limitations posed by transferring our knowledge of modern methods to ancient practices means that the data gathered from modern may not be a secure basis to as a guide to use working practices of Roman blacksmiths.

In order to expand my knowledge I went to work in the Blist Hill Iron Works in the Ironbridge Gorge Museum. Hands-on experience in the processes of making wrought iron gave the opportunity to gain some understanding of the working characteristics of bloom iron. Bloom iron is very fragile even when it is in a large mass, and large amounts of material fall off the mass if it is subjected to harsh treatment. The temperature of the bloom has to be high enough (yellow heat) to enable the slag to become liquid and so make it easier to expel. When the iron is first placed under the power hammer it is squeezed slowly and a large amount of slag comes out as a liquid. It is then subjected to very heavy hammering before it goes for rolling. This experience of working with bloom iron proved invaluable, for it provided a greater understanding of the material than can be gained from the literature.

In the first few experiments, the working characteristics of bloom iron, as well as the researcher's own performance, had to be carefully observed, if the nature of bloom iron was to be clearly understood. From the work of Crew (1991), it was shown that the type of furnace used for bloomsmithing was crucial to the successful processing of bloom iron. Crew (1991) encountered problems with tuyère blocking, which slowed down the time taken to smith the blooms he had made. The design of a bloomsmithing hearth needs to be considered to determine if there is a need for a special purpose furnace or not. Bloomsmithing for experimental purposes should not be equated with bloomsmithing on an industrial scale. The latter will require equipment that will work for prolonged periods with little maintenance and will be substantial enough to withstand possible harsh treatment by the operatives.

McDonnell (in Ottaway 1992, 471) has suggested that artifact production is a two stage process - the first is smelting, and the second blacksmithing. This is an oversimplification, because it implies that the refining of a bloom (bloomsmithing) is simply a blacksmithing process. It is true that superficially the two processes appear similar, in as much as they involve heating and hammering. The skill required to refine a bloom is a single skill in a very narrow and specific area, whereas blacksmithing requires many skills, covering a wide area of expertise. The skills needed to become a bloomsmith can be acquired very quickly, and a workman who could only smith blooms would be regarded as semi-skilled in today's terminology. The skills of a blacksmith take many years to learn, and a blacksmith is regarded as a skilled craftsman. It seems a more realistic approach to consider the making of artifacts as a three stage process - smelting, bloomsmithing, and blacksmithing. All these processes are separate and distinct.

Bloomsmithing is the intermediate process between raw iron from the bloomery and workable iron. It is subject to the problems of both the smelting and smithing processes. The iron being smithed has large amounts of slag present, which will act on the tuyère in the same manner as slag in a smelting furnace.

During the Roman period, iron production was on a large scale (Tylecote 1976, 53), and economy has to be considered as a factor in production. When iron making was in the hands of private enterprise (Cleere 1971), it has to be assumed that it was produced for profit. It seems reasonable to assume that the industry was run on production methods that would be the most economical on labour and materials. The price received for the finished product must be greater than the production costs. What is under investigation in the experiments is the production of iron on an industrial footing, not a small production process to supply a small population.

3.2 Review of the various designs of hearth for bloomsmithing

Bloom iron is a mixture of slag and iron and is of no use for the production of artifacts. The material has to be refined. This process has been explained above (c.f. section 2.4).

Bloomsmithing is a process requiring understanding of the material and its working properties. Bloomsmithing must be regarded as a specific specialised skill. It is possible that specialist workers performed the tasks and specialist equipment such as purpose-built furnaces were used. The "Stryng" hearth has been proposed as a possible special-purpose furnace for bloomsmithing (Wynne and Tylecote 1958, 339). It was used "for further heating the bloom to raise the viscosity of slag" (Lapsley 1899). However, the Stryng hearth is mentioned in the fifteenth century, and there is no positive proof to link it with the Roman period.

A furnace found at Woolsthorpe, in Lincolnshire, (Hannah, 1932) is proposed as a possible furnace or bloomery. Figure 3.1 shows a schematic drawing of the furnace. Fragments of ironstone with slag runs were found, including some pieces of iron. This indicates that the furnace was used in the iron-making process. Clearly it has been designed with a specific purpose in mind. The slope may be to assist the flow of slag away from the bloom and the deep dip may be a collection point. This is the only example of this sort of furnace to come to light and Hannah (1932) suggested that this may be an experimental furnace but gives no evidence to support this view.

A furnace at Wilderspool, Lancashire is described as a hearth for bloomsmithing. This is shown in cross section in Fig. 3.2. Schubert (1957) proposes that the funnel leads to a hole running to the outside of the furnace which allows liquid slag to escape. This interpretation seems unlikely as the slag would solidify rapidly in the hole and there seems to be no way to remove it from the tunnel. Both types of furnace are large and presumably are so to accommodate large blooms or large quantities of smaller blooms. In order to accommodate large pieces of iron, such as the so called "bath house beams", special furnaces were constructed (Fig. 3.3).

Figure 3.4, from a Greek vase, shows a large shaft-like structure in the centre of the picture. The top of the shaft is stopped with a stepped plug, so presumably the venting of gases is not possible, which would rule this out as a smelting furnace. The two figures represent a blacksmith and his striker, but the size of the furnace is considerably larger than a normal blacksmith's hearth and could be for a special purpose. The item held in the blacksmith's tongs is of no recognisable form for an artifact, and from the shape it could well be a bloom being processed. Crew (1991, 29) has suggested that the bases of redundant shaft furnaces were used to smith bloom iron.

The above examples show that some furnaces used in iron working were specifically designed to fulfil special needs. The time and trouble involved in building such furnaces can only be justified if more simple technology has been shown to be inadequate.
Illustration 10 of Figure 3.5. shows a very simple technology can be used to smith blooms, but this will only be practical if the demand for iron is small. When large quantities are called for, this simple approach will not be adequate to meet demands and a method must be found to increase production which will conserve heat and allow the heating of several blooms at the same time.

Bloom iron has poor thermal conductivity due to the slag inclusions, and so a considerable amount of time has to be spent in reheating it before forging. If the bloom is extracted from the furnace while it is still hot and smithed at once, there is a significant reduction in time for forging.

Bloomsmithing requires considerable space and the logistic problem of blacksmiths working in close proximity with shaft furnaces producing iron is difficult, but not impossible, to overcome. A detector to measure levels carbon monoxide fumes was installed near the furnace and this showed that fumes build up very quickly, which is detrimental to health, CO can be lethal. It is clear that buildings must have been constructed in such a way that fumes were reduced to a level where workers could perform their tasks without being overcome in less than an hour.

3.3 Bloomsmithing Hearth Designs

As the previous discussion demonstrates, there were special furnaces in use in the iron industry, but there is little certainty about which types of furnace were used to heat bloom iron. Several options are open, and a logical approach is to examine the major criteria that should govern the design of an ideal bloomsmithing hearth.

It is expected that slag on or near the surface of the bloom will be liberated during heating, and that the slag will collect in the bottom of the hearth. This slag will probably remain in a semi-liquid condition for a time, and the bloom iron coming into contact with it will be contaminated. In this context, slag is expelled by hammering, and when the bloom is put back into the hearth, it picks up more slag liberated from the bloom and collected at the bottom of the forge hearth. This is clearly not desirable, so facilities were provided to encourage slag to run out of the hearth and not contaminate the part-worked bloom iron.

A hearth for the smithing of bloom iron should be large enough to hold sufficient fuel to cover a piece of bloom iron to a depth where there will be enough heat to raise the bloom's temperature in order to forge it. There should be provision for the escape of slag so the bloom does not become re-contaminated. The wall thickness of the furnace should be sufficient to ensure the minimum of heat loss, and there should be provision for the introduction of air to the fuel. These are the minimum requirements for a bloomsmithing hearth.

The options discussed in the previous section (3.2) indicate that it is possible that the hearths were made from clay. Crew (1991, 29) has suggested that the bases of shaft furnaces were used for smithing blooms, and he designed a bloomsmithing hearth similar to the smelting furnace, but with a superstructure only 10 cm higher than the blowing hole.

Experimental Bloomsmithing Hearth No. 1

The design of the hearth found at Woolsthorpe (Hannah 1932) shown in Figure 3.2 was thought to fulfil the minimum requirements for a bloomsmithing hearth, and this was made to half size using clay. The floor sloped down and there was a pit to collect slag. The hearth (which was designated Bloom Hearth No. 1), is shown in section in Figure 3.6. It was made from clay, then left to dry for four days. At the end of this time the clay was of a leathery nature and was suitable for use.

To minimise thermal shock, the hearth was first heated gently. It was then filled with oak chips and set alight using only a very light draught. More fuel, in the form of small pieces of oak, was added at regular intervals for 1 hour 10 minutes.

The outside of the furnace showed no signs of cracking after this time, so the charcoal was added and the draught increased for a further thirty minutes. At the end of this time, the whole of the inside of the furnace was bright red. At this point, the bloom was inserted. The hearth was run for a further 1 hour 50 minutes before the experiment was concluded.

Observations

It was very difficult to heat the bloom right through and at times there were black (cold) spots on the bloom. This was because the bloom, with a volume of approximately 60% of the capacity of the forge hearth, was too large. The sides of the hearth were not high enough to allow sufficient cover of fuel above the bloom. The clay cracked very badly and some was lost; this was because to the clay lining was too thin on the walls. The tuyère was too close to the floor of the hearth, and slag from the floor of the hearth soon obstructed the tuyère so that no air would pass through it. The slag did not become liquid enough to flow away from the floor of the hearth. A large quantity of slag formed round the tuyère (Figs. 3.7 & 3.8). It can be seen that the tuyère is not completely obstructed, but that significant amounts of slag had built up. There is no indication of slag running away from the tuyère and into the pit.

When the hearth had cooled down, the slag was removed from the tuyère. This caused some surface cracking around the tuyère, which was repaired with fresh clay. The next experiment (Bloom Experiment 3) was carried out using the same hearth, but only lasted 38 minutes 4 seconds as the tuyère became partially blocked. The blockage seemed to be in the lower half of the tuyère, with the result that it redirected the air flow upwards, away from the heart of the furnace. This meant that the lower portion of the furnace cooled down and the temperature was insufficient to heat the bloom. On the following morning, the furnace was cold. It was emptied, and on inspection the tuyère was found to be almost completely blocked, and a large quantity of slag had formed on the floor of the hearth. The slag around the tuyère had to be removed so further experiments might take place. The large quantity of slag on the floor of the hearth may have contributed to a more rapid build-up of slag around the tuyère, so it was removed at the same time. The slag was removed with a cold chisel and hammer using only light blows. This was sufficient to cause severe cracking and disintegration, so rendering the furnace irreparable. It is possible that in all these operations the furnace never reached a temperature sufficient to allow the slag to become fluid enough to flow into the pit at the front of the furnace. This was because the tuyère was too close to the floor of the hearth, and air was not circulating through the furnace. This low temperature meant that the slag accumulated round the tuyère and did not flow away, eventually blocking off the tuyère. It is possible that the slag could be removed while the furnace was still hot.

Observations

Air will still pass through the tuyère, but because of the position of the slag build up, it will be directed upwards. This means that the hottest part of the fire will be towards the top, where there is more heat loss.

The previous experiment showed that this slag had to be removed if the tuyère was not to become obstructed again. The slag was removed with a single light blow using a hammer and cold chisel, the slag came away cleanly, and no repairs were necessary.

Conclusion

The blockage in the tuyère redirected the air flow into the furnace and after three hours it was running less efficiently and the blockage was nearly complete. Slag can be removed from around the tuyère when it is cold without causing damage to the hearth, but the slag on the floor of the hearth is almost impossible to remove without causing considerable damage. The size of the finished bloom (see section 3.5) was too small to be processed into an artifact, and is a size which is very difficult to handle for forging purposes. These small pieces could have been welded into a large block for forging into an artifact.

Bloom Hearth No. 2

The hearth was constructed of the same type clay as the other experiments and was rectangular in section. The tuyère was placed higher in the back wall to prevent slag from the floor blocking it. No pit was provided for the collection of slag, but a hole was made in one of the front corners of the hearth to provide an exit for liquid slag. As before, the hearth was allowed to dry out for several days before it was used for smithing.

The next bloomsmithing experiment (Bloom Experiment 4) was conducted for a running time of 38 minutes 22 seconds. The furnace was run at a higher temperature, with an attendant greater consumption of fuel. The tuyère was partly blocked with a build-up of slag , but still allowed enough air to circulate to heat the bloom. On the following day, the slag was removed by means of light taps with a cold chisel and hammer. The slag came away very cleanly, and there was little damage around the tuyère. No slag had flowed out of the hole and there was only a small quantity on the floor of the hearth. The front wall of the hearth was cracked very badly and needed a considerable amount of repair.

After repairs, the hearth was used for a further experiment (Bloom Experiment 5). On this occasion, a running time of 3 hours 3 minutes was recorded. The build-up of slag from the bloom had blocked off the tuyère, and liquid slag had run into the tuyère and had been solidified by the air passing through. The back wall with the tuyère was cracked and the other walls (which were too thin) had crumbled. The inside of the furnace was still hot twenty three hours after the end of the experiment, giving a good demonstration of the insulating properties of the clay. The slag was removed with a light tap, and it came away very cleanly with very little clay adhering to it (see Fig. 3.9). The air blast had to be stopped for the extraction of the bloom. If it was left on, sparks flew into the face of the operator, and the heat made the lifting of the bloom very uncomfortable. The hearth was repaired and a further experiment (Bloom Experiment 6) lasted only 1 hour 8 minutes before the tuyère blocked with slag from the bloom. Inspection revealed that although it was possible to remove the slag, the lowest layer had fused with that which had remained from the last experiment. The slag was deeply embedded in the clay and it was apparent that a build-up was taking place. The sides of the hearth almost crumbled away, and it was decided not to rebuild them. The front was repaired with fire brick and loose sand. The next experiment (Bloom Experiment 7) lasted only 49 minutes before the tuyère blocked, and it was decided that this hearth was no longer suitable for repair.

Bloom Hearth No. 3

A new hearth was built to try to overcome some of the difficulties of the previous hearth. The walls were made thicker, and the tuyère was made to slope down, so directing the air towards the floor of the hearth. For the next experiment (see Bloom Experiment 8), the objectives were changed. The effects of heating a bloom for an extended period at high temperature were examined in order to see if iron would separate from slag, giving a bloom which would be more readily forgeable. Charcoal found on excavated Roman ironmaking sites was about 25 mm cube Cleere (1971). This is charcoal that would have been used in smelting. The author's experience of using charcoal for forging has shown that it is unnecessary to have charcoal of uniform size. Pieces that range from 5 mm cube up to 40 mm cubes are suitable. It has been found that a charcoal fire is easier to start if the pieces of charcoal are small so the hypothesis that only a small amount of charcoal needs to be cut to small pieces at the start of the smithing was examined.

Observations

The furnace was allowed to cool, and on removal of the ash, the bloom was found to have formed into a large flat cake. The material had adhered to the side and base of the furnace and considerable undercutting had taken place under the tuyère. The material was dark grey, and was probably iron oxide. A strong oxidizing atmosphere was present rather than the reducing atmosphere which required. When attempts were made to remove the material, the furnace split down its length (Fig. 3.10). The floor was too thin, and not enough time had been allowed for the clay to dry.

The angled tuyère directed the heat to the part of the fire where it was most needed, and observation during the experiment showed that the heat was in the fire, while the upper part stayed cooler. When the tuyère was horizontal, the whole of the fuel became hot, and in this case, any fuel placed on top of the furnace became red hot almost at once. Since the upper portion was not glowing, it was possible to put large pieces of charcoal on top of the fire. These dried out and became very brittle, allowing the operator to break them into small pieces approximately 25 mm square. It was therefore unnecessary to cut a large amount of charcoal into small pieces before the experiment started.

Conclusions

The bloom had become a mass of iron oxide, which had bonded to the floor of the furnace and could only be removed with considerable force, bringing large pieces of clay away with it. Iron oxide indicates that an oxidizing atmosphere was present, rather than the required reducing atmosphere. The experiment confirmed that large pieces of charcoal could be placed on top of the fire and then broken up with tongs. Therefore time does not need to be spent in either selecting or cutting charcoal to the correct size.

General overview

All the hearths suffered from several common deficiencies. They were all high sided, and it was necessary to stand over the top of them to extract and replace the bloom. This meant that the full heat of the furnace was directed at the operator's face. There was a large surface area of heat, and it was difficult to manipulate the bloom into the correct position in the fire, and even more difficult to extract it as it became smaller. The longest running time was just over three hours. In many cases the blooms were not fully smithed as the hearth became inoperable, due to tuyère blockage. In most cases, considerable damage was done when the slag adhering to the tuyère was removed, often causing damage which was beyond repair.

Bloom Hearth No. 2 ran the most efficiently and sustained the least damage when slag was removed. The experiments show that a high-sided container made from clay, with provision for air to be introduced near to the base and below the level of the top of the fuel, is a most unsatisfactory form of design for a bloomsmithing hearth.

3.4 Tuyère experiments

Bloomsmithing experiments had not proceeded at all well, reflecting the unsatisfactory design of the hearth. One of the major problems was the blocking of the tuyère. It was decided to isolate this problem and to concentrate on finding methods of eradicating it.

The purpose of the tuyère is to conduct air from the bellows to the fire, and to protect the bellows from the damaging effect of heat. In order to protect the bellows, the tuyère is made from a material which is a poor conductor of heat, usually clay (Tylecote 1987, 115).

In the previous experiments, a thick wall of clay was built to provide a tuyère. This may be misleading, and does seem extravagant. Mongait (1961, 138) reported the find of some smith's tuyères at Kalinovka, (in the former USSR) these were from the period 1200-1400 B.C. Although from a from a bronze worker's kit, the find shows the principle of a tuyère which was manufactured independently of the furnace (Fig 3.11)

Material for tuyères

Modern tuyères for blacksmiths' forges are made from cast iron. Tylecote (1987, 327) suggests that cast iron was not produced deliberately in Europe until the thirteenth century, and any cast iron from the Roman period is either an accident in the smelting process, or comes from imports from the Far East. Therefore it could not have been used for the manufacture of tuyères. A tuyère made from wrought iron would melt in the fire and would probably seal itself. If this material had been used in antiquity, it is probable that at least some traces would have been found. Since no claims have been made for wrought iron tuyères, we must conclude it is unlikely that they existed.

The properties of clay make it an ideal material to manufacture tuyères from: clay is easy to obtain - in many areas of Britain it can be obtained locally; it is cheap; it can be moulded very quickly; and it has high heat resistance. It would be very easy for a blacksmith to make his own tuyères and use his own hearth to fire them. If a smith was to make his own tuyères, he would almost certainly make a mould to speed up the process.

Various types of tuyère are known (Tylecote 1987, 124), although most are associated with smelting or blacksmithing rather than bloomsmithing. After a careful study of the tuyères shown in Figure 3.11, the shape of the tuyère was reduced to its simplest geometric form (Fig. 3.12). The conical section of the inside hole was probably present to accept the nozzle of the bellows and would create a sound seal; however it was thought to be unnecessary for the experiments in hand.

The outside shape is a frustum of a cone. The hole down its centre has parallel sides. The outside shape is the approximate shape of a plant pot. For convenience, a plant pot was used as a mould to make a series of experimental tuyères. The same mould was used for all the tuyères in order to ensure that the volume of clay was the same. Another unknown was the size of the hole through which the air passes. The smith's tuyères found at Kalinovka, USSR (Mongait 1961, 138) had a hole diameter of 25 mm to 30 mm. A tuyère with a diameter of 27 mm was used for the first experiment.

Tuyère Experiment 1

Local clay from Tilehurst, Reading was puddled by hand, and impurities were removed. A ceramic plant pot was used as a mould and the clay was put into the mould in small pieces and pressed firmly down. A piece of copper pipe 25 mm in diameter was used to cut a hole through the centre of the clay. The clay still in the mould was left next to the flue of a central heating system for two days, in order to dry out. At the end of this time the tuyère had cracked down its length and was irreparable.

Tuyère Experiment 2

The same procedure as Tuyère Experiment 1 was repeated, but this time the mould was left in the forge. The weather was very cold and the tuyère cracked in the same manner. All these problems could probably have been avoided if a potter had been consulted first.

Tuyère Experiment 3

No mould was used to make the tuyère for this experiment. The clay was formed into a truncated cone by hand. This was left in a warm room but not close to any heat source. After two days it was hard enough to handle. The tuyère was left to dry for eleven days.

If the blacksmith made his own tuyères, he might well have made use of his own forge fire to bake them once they were dry. They could have been baked in the fire at the end of the day, and this would not have disturbed his working routine. This theory was put to the test. The tuyère was placed in the fire and charcoal was heaped over it. The fire had been in use for six hours and had been cleaned of slag, no forced draught was used. The tuyère was extracted after one and a half hours. Some baking had taken place and the clay was a dark grey colour, and, when tapped lightly with an iron bar, made a ringing sound, indicating some hardening had taken place.

Tuyère Experiment 4

A tuyère which had been made in the plant pot mould was left to dry for one week. It was then put in the forge fire at the end of a working day, and was placed with the smaller end in the hottest part of the fire, with charcoal piled up on it. It was left overnight. The next morning it was baked to a brick red at the bottom, going to a light grey in the portion furthest from the fire. The tuyère was hard and easy to handle.

Tuyère Experiment 5

During the hearth-making experiments, no temper was added to the clay, which may have been the reason why so much cracking took place. For this experiment, a further quantity of clay was puddled and then mixed with chopped straw to act as a temper in order to see if this would overcome the tendency for the clay to crack on the surface. It was baked and showed only slight cracking on the surface.

Forging using the tuyères

The tuyère made in Tuyère Experiment 5 was used in the forge fire when a copy of a Roman blacksmith's hammer from Silchester was made (see Chapter 4.4). The fuel used was charcoal. For the first 35 to 40 minutes the fire burned as normal, but after this time the position of the heart, or the hottest part, of the fire started to change place. The heart of the fire is usually directly in front of the mouth of the tuyère. After a further 20 minutes, the position stabilised. It became necessary to continue increasing the volume of air in order to keep the fire at a constant temperature, and the position of the heart again started to move. After 1 hour 25 minutes the fire started to die, and it was obvious that the tuyère was blocked. It was removed and the hammer forging experiment was continued in a different hearth. Large quantities of slag had formed on the tuyère, (Fig. 3.13) and although the tuyère was not completely blocked, the slag had caused the air to flow at right angles to the axis of the tuyère. The constant moving of the air flow made forging very difficult, because it was impossible to see clearly exactly where the hottest part of the fire was.

The tuyère made in Tuyère Experiment 3 was used in an experiment to forge raw bloom iron. It behaved in the same manner as the previous tuyère and only lasted 1 hour 12 minutes before blocking took place. On examination, the direction of air flow had been turned through 90° again.

Discussion

The separation of the tuyère from the main body of the forge hearth has not overcome the problem of the tuyère blocking. The overall design of the hearth does not influence the performance of the tuyère significantly, but the material being smithed does. Gravity causes the residues released from the fuel and the bloom to fall down into the position occupied by the tuyère, and because this is the hottest part of the fire, these residues become more viscous and adhere to it. In both the experiments, the tuyères were placed into an unlined sand pit, so no other clay was present. The high temperature caused the tuyère to melt. The melting clay combined with the slag caused the build-up. Separate tuyères do not appear to offer a viable alternative which will overcome the problem of the tuyère blocking after a short time. These tuyères were in direct contact with the fire - indeed, the hottest part of the fire was directly in front of the tuyère a short distance from the face of the tuyère. It is possible that the tuyère does not need to be in direct contact with the fire. Tylecote (1987,123) asserts that tuyères for blacksmithing last a long time, but this is not the case when the tuyère is in direct contact with the fire. The use of blowpipes in metal working is well attested in ancient Egypt (Scheel 1989, 32). Figure 3.28 shows the use of a blowpipe in the Roman period. There is no reason to suppose that the development of tuyères which were built into the wall of a furnace, or forge hearth, made the use of blowpipes obsolete in metalworking.

3.5 Experiments of bloomsmithing by accepted methods and the results

The techniques of bloomsmithing were fully explained in Chapter 2. The main objective of bloomsmithing is to expel slag and cause the iron to weld together. An examination of a piece of bloom iron showed that it was made up of small pieces of iron separated by voids and entrapped slag. The bloom is held together by the slag, and if the bloom is subjected to a heavy blow it will shatter.

It is not known exactly what form quality control took in the Roman iron industry, or if there were quality standards for Roman iron. For example, chain has to be made from iron of a high standard or it will fail. Trial and error, in the case of many iron items, is not an acceptable alternative. Crew (1994) has suggested the Iron Age currency bars are a form of quality control. If this is so, then it is possible that a similar standard could have existed in the Roman world, but its form is uncertain.

For the purposes of future experiments it was necessary to set a standard. The minimum standard of acceptability decided upon was equivalent to the lowest standard of iron produced at Ironbridge. This is known as 'muck bar' (see glossary). Figure 3.14 shows a section through a piece 'muck bar'. The highest standard will be equal to five rolled wrought iron (see glossary). Figure 3.15 shows a section through iron of this quality. A section was taken through a piece of Roman iron found during the excavations at Silchester in 1977, and marked as iron bar (ref. 001. 48 FE13 C89/81). The original function of this piece of iron is unknown. It can be seen from the cross-section (Fig 3.16) that, from the ratio of slag to iron, the Roman sample lies between that estimated in Fig. 3.14 and that in Fig. 3.15. Any piece of iron with a quality that comes between that in Fig. 3.14 ('muck bar') and Fig. 3.15 (five-rolled wrought iron) will be considered as acceptable. The performance of the experimental hearth used in the following experiments has already been discussed, in sections 3.3 and 3.4 of this Chapter.

Bloom Experiment 1

The first experiment had two objectives. The first was to test the theory that the bloom should be squeezed first, before any hammering is carried out. This theory was inspired by the observations made by the author when participating in the manufacture of wrought iron (see Section 3.1) The second hypothesis was that the shock waves created by a steel hammer are too severe, and that the bloom could be hammered into shape with a soft hammer made from lead.

Equipment: Modern forge hearth, Fuel: Coke. Lead hammer, weight 3kg, sledge hammer, weight 5kg, Tongs.

The sequence of operations for Bloom Experiment 1 is given in the table 3.1 and the results of the experiment in table 3.2:

Heat No.	Time (minutes)	Tools	Comments
1	19.0	lead hammer	Light tap. 50% of bloom fragmented
2	18.0	tongs	Bloom placed next to fire and squeezed to shape. Not very effective
3	9.0	lead hammer	Hammer fell under its own, lead stuck to bloom
4 to 12	45.0	5 kg sledge hammer	Hammer fell under its own weight, bloom brought to rough cube shape

Table 3.1 Sequence and duration of operations in Bloom Experiment 1.

Results

Starting weight of bloom	311.9 g
Finished weight of bloom	86.0 g
Size of finished bloom	65 mm x 16 mm x 15.4 mm
Material loss	72.4%
Total time	1 hour 31 minutes.

Table 3.2 Results from Bloom Experiment 1

Observations

If a lead mallet is used, large amounts of lead fuse onto the surface of the bloom. A large amount of clinker built up in the fire. It was a sticky black treacle-like substance, which became like glass on cooling.

Conclusions

There were too many variables in this experiment posing too many questions at once. It was clearly demonstrated that a lead mallet will contaminate the bloom and, as there is no evidence showing lead incorporated in blooms, we can conclude that this was not an approach that was undertaken in antiquity. It was also shown that the bloom is in a very fragile condition when it emerges from the hearth after the first heating, and great care must be exercised in handling it, if large amounts of bloom are not to be lost. Squeezing the bloom while it is at a high temperature helps to reduce material loss. There was no difficulty in heating the bloom through, however, we must remember that it was processed with modern equipment.

The sample from BEX 2 was sectioned and one cut surface was polished and etched (Fig. 3.17). This revealed large areas of slag entrapment and voids. The sample was still below the minimum standard and was unsuitable for processing into an artifact. It would need more refining before it become suitable for use. Considering the loss of material incurred, it seems likely that almost the same amount of material would be lost in order to bring the sample to a satisfactory state of refinement. Considering the small size of the sample, and the fact that it was approaching the size when it became too small to handle, no further work was carried out.

Bloom Experiment 2

Some experience was gained from the first experiment. As it was observed that large amounts of metal came off during squeezing, it was decided to attempt bloomsmithing using only a hammer (see section 3.3: Bloom Hearth No.1)

The objective of this experiment was to test the effect of using only hammer blows to consolidate a bloom.

Equipment: Experimental Bloom Hearth No.1; Fuel, charcoal; Block anvil; Steel-headed hammer, weight 2.5 kg; Tongs. Bloom weight 575.0g.

The sequence of operations for Bloom Experiment 2 is given below in table 3.3 and the results in table 3.4:

Heat No.	Time (hr : min)	Tools	Comments
1	1 : 10	Bloom hearth No. 1	Furnace filled with oak chips, using only a very light draught; fuel added at regular intervals.
2	0 : 30	Bloom hearth No. 1	Charcoal was added and the draught increased
3	0 : 08	Hammer, anvil, tongs.	Bloom removed; light hammer blows; bloom fell in half.
4	0 : 09	Hammer anvil, tongs	Light blow, bloom fell into small pieces.
5 to 10	1 : 03	Hammer, anvil, tongs	Bloom had broken up into a number of pieces.

Table 3.3 Sequence and duration of operations in Bloom Experiment 2

Results

Starting weight of bloom	575.0 g
Finished weight of bloom	143.0 g
Fuel consumption	2.5 kg (of charcoal)
Material loss	75.1%
Total time (hammering time only, preheating time not included)	1 hour 50 minutes

Table 3.4 Results from Bloom Experiment 2

Conclusions

If the bloom is only hammered in the first instance, very large amounts of the bloom are lost, even when the temperature was high and no heavy blows were struck with the hammer. The pieces that fall off are too small to be processed on their own. The blocking of the tuyère may be the reason that the forge did not get as hot as it should. The operator suffered nausea and headache for several hours after the conclusion of the experiment. This was probably carbon monoxide poisoning.

The sample produced in bloom experiment 2 was sectioned and one cut surface was polished and etched (Fig. 3.18). This revealed large areas of slag entrapment and voids, but these were smaller and more even than in the sample shown in Figure 3.17. However, it was still below the minimum standard and was unsuitable for processing into an artifact. It needed more refining before it was suitable for use. The loss of material incurred at the termination of the experiment was 75.1%. It is probably likely that almost as much material will be lost again before the sample can be brought to a state of refinement equal to the lowest standard acceptable for these experiments. Figure 3.14 shows the minimum standard.

Bloom Experiment 3

In Bloom Experiment 1, a combination of squeezing and hammering was employed but this was found to have two major faults. The soft hammer contaminated the iron, and the tongs were too short to provide the leverage needed to exert a large enough force to squeeze the bloom. Since it was felt that squeezing the bloom first would reduce material loss, Bloom Experiment 1 was modified, and a steel hammer was used in combination with the largest pair of tongs available.

Equipment: Bloom Hearth No. 1; Fuel: charcoal; Block anvil; Steel headed hammer, weight 2.5 kg; Tongs.

Objectives: First, to observe the effects of a combination of hammering and the use of tongs to consolidate a bloom. Secondly, to observe the effects when the same hearth was used, and a smaller bloom was processed.

The sequence of operations for Bloom Experiment 3 is given below. in table 3.5 and the results in table 3.6.

Heat No.	Time (mins : sec)	Tools	Comments
1	7 : 15	Hammer	Five light blows
2	12 : 03	Hammer	Heavy blow, fell in half
3	10 : 35	Hammer	Heavy blow, fell in half
4	3 : 01	Tongs	Easy to squeeze, no loss
5	6 : 00	Tongs	Fell to pieces, tuyère blocked

Table 3.5 Sequence and duration of operations in Bloom Experiment 3

The forge hearth was fired up at 11:15 hours, and was up to working heat in 13 minutes. At 12:06 hours the experiment was terminated.

Results

Starting weight of bloom	203.0 g
Finished weight of bloom	46.0 g
Fuel consumption	4 kg (of charcoal)
Material loss	72.9%
Total time (hammering only preheating not included)	52 minutes.

Table 3.6 Results from Bloom Experiment 3

Observations

In spite of the duration of heat number 5, the bloom did not get any hotter, because of the partial blockage of the tuyère. Figure 3.14 shows the blocking of the tuyère. The blockage was in the lower half of the tuyère hole, and this directed the air flow in an upwards direction, away from the heart of the furnace. The top half of the hearth was hot, but the bottom was almost cold. In order to heat the bloom right through, it had to be covered with hot fuel. When this was done, the hottest part of the fire was above the bloom, rather than underneath it. When the bloom was extracted, only the surface was at the required temperature, and the heat had not penetrated the mass of the bloom. Hence, when pressure was applied, the bloom disintegrated. The operator experienced nausea and headache for several hours after the experiment.

Conclusions

After a running time of 3 hours, the tuyère had become less efficient. The slag build-up directed the air flow in an upwards direction. This change in air direction shows that the tuyère was starting to become blocked. In this experiment slag was removed easily, with little damage to the hearth, when the hearth was cold. The size of the finished bloom was too small to be forged into an artifact. The sample was sectioned and one cut-surface was polished and etched Fig. 3.19. The polished section revealed large areas of slag entrapment and voids. There is an improvement in quality, because less slag is showing up on the section, but the sample is still below the minimum standard and was unsuitable for processing into an artifact. It needed more refining before it was suitable for use.

Observations on Bloom Experiments 1, 2 and 3

The blocking of the tuyère was an unforeseen obstacle, and could mean that the design of the hearth would have to be modified. There had been no definite course of action concerning the use of tongs and hammering, so the next experiment would concentrate on using only tongs until the bloom would not contract any further. An understanding of the way bloom iron behaves was beginning to emerge. High heat alone did not ensure that a bloom would weld, and long heats caused cracks to develop which were not apparent to the naked eye. So far, although only small pieces had been produced it was possible that these could be welded into a larger block for forging into an artifact.

Bloom Experiment 4

In this experiment the bloom was squeezed with tongs until it would not compress any further.

Equipment: Bloom Hearth No. 2; Fuel: charcoal; Block anvil; Steel headed hammer weight 2.5 kg; Tongs.

The sequence of operations for Bloom Experiment 4 is given below in table 3.7 and the results in table 3.8

Heat No.	Time (min : sec)	Tools	Comments
1	9 : 14	Tongs	Squeeze, a lot of resistance
2	1 : 15	Tongs	Easy to squeeze, bloom made noise like frying bacon
3	5 : 26	Tongs	Squeeze easily into round section, noise like frying bacon from bloom
4	5 : 29	Tongs	As 3 but no noise
5	2 : 18	Tongs & hammer	Hard to squeeze, light hammer blows
6	2 : 15	Tongs & hammer	Hard to hammer, fire coming out of drain hole
7	3 : 20	Tongs & hammer	Bloom parking hammer, forming cube shape
8	3 : 14	Hammer	Now possible to work at lower temp. forming cube shape
9	4 : 10	Hammer	Rectangular shape forming
10	2 : 42	Hammer	Rectangular shape forming
11	3 : 53	Hammer	Rectangular shape forming
12	3 : 47.	Hammer	Rectangular shape forming
13	3 : 59	Hammer	Rectangular shape forming
14	3 : 59	Hammer	Rectangular shape forming.
15	3 : 36	Hammer	Rectangular shape forming.
16	3 : 54	Hammer	Rectangular shape forming
17	3 : 12	Hammer	Rectangular shape forming
18	3 : 57	Hammer	Rectangular shape forming

Table 3.7 Sequence and duration of operations in Bloom Experiment 4

Results

Starting weight of bloom	249.0 g
Finished weight of bloom	61.0 g
Fuel consumption	4 kg (of charcoal)
Material loss	75.5%
Total time (hammering time only preheating time not included)	1 hour 15 minutes

Table 3.8 Results from Bloom Experiment 4

Observations

After 31 minutes 41 seconds (heat No. 8), forging became easier and it was possible to forge the bloom into a billet using only a hammer. The bloom was easier to squeeze initially; this was because the furnace was run at a higher temperature of 1000 °C, with the attendant greater consumption of fuel. There was a noticeable reduction in production time when smithing the bloom. The sample was sectioned and one cut surface was polished and etched (Fig. 3.20). This revealed that the areas of slag entrapment and voids had been reduced considerably. The sample was still below the minimum standard and was unsuitable for processing into an artifact. The total loss of material at the end of the experiment was 75.50%. It seems likely that almost the same percentage loss of material would occur again before the sample would be brought to a state of refinement equal to the lowest standard acceptable for these experiments . This is shown in (Fig. 3.14) which is a section through a piece of "muck bar".

Conclusion

This experiment has shown that a consistent high temperature 1000° C to 1100°C is necessary for bloomsmithing. For the production of large volumes of iron, a furnace with a tuyère which is consistently blocking will reduce production to an unacceptably low level. The other principal difficulty is trying to keep the bloom in one piece in order to keep it at a suitable size to forge with a hammer.

Bloom Experiment 5

Evenstad (1968) has stated that a flux made of sand and ground-up hammer scale was used in bloomsmithing. It was hoped that this flux would assist in cleaning the iron and encouraging the slag to flow. To test this hypothesis, an experiment was conducted to observe the effects of a flux. Slag, which is liberated either by heat or hammering, may leave behind traces on the surfaces which it has just vacated, which will inhibit welding. It is possible that flux will help to dissolve any slag film and make welding easier.

Equipment: Bloom hearth No 2. Fuel, charcoal; Block anvil; Steel headed hammer (weight 2.5 kg); Tongs.

Objective: To observe the effects of using silver sand as a flux during initial forging of a bloom.
Furnace pre-heated for 11 minutes.

The sequence of operations for Bloom Experiment 5 is given below in table 3.9 and the results in table 3.10.

Heat No.	Time (mins : secs)	Tools	Comments
1	11 : 01	Hammer, silver sand.	1/3 of bloom fell off in the fire
2	4 : 53	Tongs, silver sand	Hard to squeeze.
3	7 : 47	Tongs, silver sand	Bloom deep in fire to make hotter. Removed with rake.
4	8 : 37	Hammer	Light blows, little change in shape
5	9 : 23	Hammer	Light blows, little change in shape
6	10 : 42	Hammer	Light blows, split developing
7	11 : 53	Silver sand	Silver sand in split to act as flux for welding.
8	17 : 11	Hammer	Blows too heavy, split section fell off, iron not hot enough
9	4 : 05	Hammer	Moderate blows, some change in shape
10	4 : 01	Hammer	Getting iron hotter, heavier blows
11	4 : 0 3	Tongs	Squeeze into cylinder shape.
12	3 : 11	Hammer & tongs	Squeeze into cylinder shape
13	4 : 03	Hammer & tongs	Hard to squeeze light hammer blows.
14	4 : 03	Hammer & tongs	Near welding heat heavy hammer blows
15	4 : 01	Hammer	Near welding heat heavy hammer blows
16	3 : 07	Hammer	Large pieces fell off fire dying possible tuyère blocking?
17	5 : 21	Hammer	Fire dying, increasing blast does not help.
18	4 : 00		Fire temperature too low for forging. Bloom lost in fire.

Table 3.9 Sequence and duration of operations in Bloom Experiment 5

Results

Starting weight of bloom	671.0 g
Finished weight of bloom	170.0 g
Fuel consumption	3 kg (of charcoal)
Material loss	74.7%
Total time (hammering time only preheating time not included)	1 hour 15 minutes

Table 3.10 Results from Bloom Experiment 5

Comments

The build-up of slag inside the furnace had blocked off the tuyère and liquid slag had run into it. This had been solidified by the air passing through the tuyère. The back wall with the tuyère was cracked, and the other walls were thin and crumbling. The bloom was too small to be forged into an artifact, and got lost in the fire. The great weight loss showed this is an inefficient method of production and the time consumed indicated a high use of both man-hours and fuel. Twenty six hours after the start of the experiment the inside of the furnace was still warm. The slag was removed with a light tap, coming away very cleanly with very little clay adhering to it. No repairs were necessary. At this stage, silver sand as a flux seems to make very little improvement in the forgeability of the bloom. The air blast had to be stopped for the extraction of the bloom. If it was left on sparks flew up into the face of the operator, and the heat made the lifting of the bloom very uncomfortable. The sample was sectioned and one cut surface was polished and etched (Fig. 3.21) This revealed large areas of slag entrapment and voids. The sample was still below the minimum standard and was unsuitable for further processing.

Observations on the experiments using Bloom Hearth No. 2

The design of the hearth would have to be altered. The most important requirement for the bloom iron, was the need to keep it together long enough to start hammering. It was thought possible that soaking the bloom in a high temperature for a prolonged period would cause the slag to run out and allow the cleaner iron to adhere, so making it possible to refine a bloom in a shorter time.

The furnace was allowed to cool, and on removal of the ash the blooms were found to have gone to a large flat cake. The material adhered to the base and sides of the furnace and considerable undercutting had taken place by the tuyère. The dark grey material was iron oxide. A strong oxidizing atmosphere had been created, whereas a strong reducing atmosphere was required. When attempts were made to remove the material, the furnace split down its length. The floor was too thin and not enough time had been allowed for the clay to dry. The angled tuyère seemed to direct the heat to the portion of the fire where it was most needed and observation during the experiment showed that the heat was low in the fire while the upper part of the fire stayed cooler. When the tuyère was horizontal the whole of the fuel became hot. In this case, any fuel placed on top of the furnace became red hot almost at once. However, in this experiment, as the upper portion was not glowing, it was possible to put large pieces of charcoal on top of the fire; these dried out and became very brittle, allowing the operator to break them up into small pieces, 30 mm cubed using tongs. It was therefore unnecessary to cut a large amount of charcoal into small pieces before the experiment started.

Conclusions

The blooms had become a mass of what appeared to be iron oxide, which was bonded to the floor of the furnace and could only be removed with considerable force, bringing away large pieces of clay with it. The iron oxide indicated that the furnace had a strongly oxidising atmosphere when a reducing atmosphere was required. Large pieces of charcoal could be placed on the top of the fire and after a suitable time could easily be broken up into a size suitable for burning in the hearth, i.e. about 30 mm cubes.

Bloom Experiment 9

Previous experiments were conducted in purpose-built forge hearths. These were time-consuming to build and, because of their design, made extraction of the bloom extremely difficult. The operator was burnt from sparks, and became ill from carbon monoxide fumes. All the designs encountered problems with the tuyère blocking. In view of these difficulties a new approach was decided on. The most pressing problem was finding a suitable method of extraction of the carbon monoxide. It was decided to make use of the existing forge hearth at Ironstone Forge (see chapter 5). It had good ventilation for fumes through the chimney and was at a suitable height for comfortable working. It had a cast iron tuyère and electrically driven fan-bellows.

In previous bloom smithing experiments, the tuyère was simply a hole in a wall of clay. When this became blocked with slag, the experiment had to be terminated. It was possible to free some of the obstruction by pushing a wet twig in the tuyère hole from the back. Once the slag became hard no air penetrated and the experiment had to stop. This is an unsatisfactory position as it was impossible to complete any experiment without first having to close the forge hearth down completely for repairs. It must have been impossible for early iron makers to produce iron in this manner. So a different approach had to be adopted: small tuyères are known (Tylecote 1987, also see section 3.4 of this chapter) so it was decided to make small clay tuyères that would fit over the end of the cast iron tuyère. The tuyères were made from Tilehurst clay with various tempers (section 3.4).

The objectives of this experiment were:

(a) To test the effectiveness of clay tuyères;
(b) To use a bloomsmithing hearth with a sand floor, to see the resulting slag formation and compare it with the archaeological record; and,
(c) To use a flux of silver sand and hammer scale.

The weight of the raw bloom was 1385g. Tongs were used for the first 5 heats and flux was applied before the bloom was returned to the fire. The flux seemed to help to hold the bloom together and produced a stickiness on the bloom. After 51 minutes 19 seconds, the tuyère was blocked. It was simple to remove it with a pair of tongs and replace it with a fresh one. The whole operation took less than two minutes. After heat 6 a light blow was struck and the bloom fell in half. It became apparent that the main cause of the disintegration of the bloom was that it was possible for it to move in too many directions at the same time. Thus if it was possible to restrain the movement of the bloom, the necessary welding of the iron would be easier.

A piece of the bloom was heated as before and fluxed. This dropped down a piece of square steel tube (75 mm inside dimension and 100 mm long) that was held with tongs on the anvil. A square steel bar 73 mm square and 120 mm long was put into the tube on top of the bloom and struck heavy blows with a 5 kg sledge hammer. The resulting slug of iron appeared to be a cohesive mass. The procedure was repeated for the remaining pieces of iron and the results were the same. One of these pieces was sectioned (Fig. 3.22). It has some slag inclusions, but it is of a higher quality than the iron produced by hand methods.

Discussion

A considerable amount of slag built up on the face of the tuyère and sealed it off. The tuyère was left for three minutes and was lifted from the hearth with the slag adhering to it. This was a very simple and quick operation. The second tuyère was run for 49 minutes and the slag was removed in the forge while the tuyère was still hot. No clay came away with the slag and the tuyère was used in the next experiment. This method shows that it is possible to process iron in a normal blacksmith's hearth and that the problems of tuyère blockage can be overcome by replacement of tuyères. Flux, the mixture of silver sand and hammer scale, improves the process of smithing by holding the bloom together during hammering. It encourages the flow of slag from the bloom. The combined slag and flux may account for the 'sticky' feel of the bloom during the initial forging.

Enclosing the bloom

Enclosing the bloom during hammering does speed up the time necessary to produce good quality iron. 502 g of iron were produced in 1 hr. 40 mins. This represents a considerable improvement on previous experiments. In the next experiment, the bloom was cut into small pieces and the enclosing system used from the start, employing tongs, rather than a tube.

Results and objectives of bloomsmithing experiments by conventional methods

Bloom Experiment 1

(a) Tests whether the bloom should be squeezed first, before hammering.
(b) Tests possible use of a soft hammer such as lead.

Bloom Experiment 2

To test the effect of using only hammer blows to consolidate a bloom.

Bloom Experiment 3

Observes the effects of a combination of hammering and the use of tongs to consolidate a bloom.

Bloom Experiment 4

Starts by squeezing bloom with tongs - continues until it will not squeeze any more.

Bloom Experiment 5

Observes the effects of using silver sand as a flux during initial forging of a bloom.

Bloom Experiment 8

Examines the effects of heating a bloom for an extended period at a high temperature to see if slag would separate from iron, giving a more readily forgeable bloom.

Bloom Experiment 9

(a) Tests the effectiveness of clay tuyères.
(b) Uses a bloomsmithing hearth with a sand floor, to see the slag formation and compare it with the archaeological record.
(c) To use a flux of silver sand and hammer scale.

Summary of Results

EXP. No.	Initial weight of bloom (g)	Finished weight of bloom (g)	% of material lost	Forging time hour : min
EXP 1	312	86	72.4 %	1 : 31
EXP 2	575	143	75.1 %	1 : 50
EXP 3	203	55	72.9 %	0 : 52
EXP 4	249	61	75.5 %	1 : 15
EXP 5	671	170	74.7 %	3 : 03
EXP 6	Fan failed. No iron produced.			
EXP 7	No iron produced			
EXP 8	Iron oxide produced			
EXP 9	1385	502	63.75 %	0 : 52

Table 3.11 Summary of yields and production time in Bloom Experiment 1 - 9

EXP No.	Production time/gramme (minutes)
EXP 1	1.06
EXP 2	0.77
EXP 3	1.06
EXP 4	1.23
EXP 5	1.08
EXP 9	0.10

Table 3.12 Production time / gramme of iron for Bloom Experiments 1-5 and 9

Summary

Experiments so far have shown that the smithing of a single bloom is very time-consuming. The results of Bloom Experiment 9 have shown that fluxing decreases the production time; but as there is no positive proof that fluxing was used in bloom smithing Bloom Experiment 9 has not been used in the calculations for bloomsmithing times. The average production time for bloomsmithing a single bloom taken from EXP 1-5 is 1.04 minutes per gramme or 17 333 hours per tonne.

3.6 Justification for a different approach

The previous experiments were based on the theory that bloom iron has to be refined as a single bloom, and gave a production time of 17 333 hours per tonne. This translates to a production time for bloomsmithing of *39 000 000 man hours per year* if Aiano's (1975) figure of 2 250 tonnes per annum is accurate for Roman Britain.

Experiments have shown that the slag from the bloom and the melting clay caused the tuyère hole to block. Unblocking the hole usually inflicted substantial damage to the forge hearth, often rendering it unfit for further work. Such a short working time for a furnace would make production almost impossible. Experiments have failed to show how these problems could be overcome using the currently accepted views of bloomsmithing.

Bloomsmithing a different approach

Iron was produced to meet the needs of both the civil and military establishments. The supply of weapons for the Roman army demonstrates the need for a continual supply of iron. Experiments to find possible ways of making workable iron have to be considered within the context of how much iron was produced, who was the user, what was the time scale for production, and how much manpower was required. It has already been shown that if the current model is followed, then the manpower required is enormous.

If bloom iron is to be refined by working on a large bloom, the time necessary for iron production on the scale required by the Roman imperial armies would be unreasonable with these methods. When this is combined with the problems of the tuyère blocking and the hold-up this causes, this seems to be an unworkable method of refining iron, and a different approach to the problem is required.

There are two different and distinct problems,
(i) the bloom breaking up; and,
(ii) the tuyère blocking.

In order to try to arrive at a more realistic solution, they should be considered and analysed separately.

The breaking up of the bloom

A body, subjected to force, will break along the line of least resistance. In a bloom of iron, the points of least resistance will be where the slag inclusions are largest, or where there are other defects in the structure, such as voids. Crew (1991) based his bloomsmithing experiments on the assumption that blooms were smithed as a whole. There is evidence to show that this was not always the case. Iron billets from the Roman forts of Strageath and Newstead have been examined by the author. Figure 3.23 shows a billet from Newstead number (FRA 325). The weld lines can be seen quite clearly and shows the various small pieces that make up the whole billet. The billet has three large pieces forming the majority of the piece, but there are also several smaller pieces incorporated on the left side.

Although from a later period, billets from Haithabu in North Germany, from the 9th to 11th centuries were made from small pieces of iron welded together (Thomsen et al. 1971, 22-23; see also Figure 3.24).

Agricola (1950, 423) makes reference to blooms being cut up into small pieces for forging. In more recent times, Busch (1972) describes blooms being extracted from the furnace and being immediately cut up at red heat. On the strength of this evidence it was decided to adopt this practice as the model for further experiments. Bloom iron was obtained from Blist Hill Iron Works, Ironbridge. It was bloom iron that had been reduced, but not passed under the drop hammer or through rollers. The condition was probably slightly superior to Roman bloom iron in that it might have had slightly fewer slag inclusions (Gale, pers. comm. 1991). During the initial stages of bloomsmithing, the greatest problem was to hold the bloom together until it was solid enough to be given sufficiently heavy blows to weld the iron and expel the slag. The bloom was hammered all over the outside with gentle blows. The outside of the bloom became consolidated iron, which effectively acted as a barrier to prevent slag from being expelled. Therefore this system was self-defeating. The bloom had a larger surface area than the hammer, and as the hammer compresses the centre of the bloom the outside surfaces have no force on them, and they are pushed at 90° to the force of the hammer (Fig. 3.25).

Tong types

The bloom needed to be contained around its sides when the hammer blow fell and the face of the hammer needs to be similar in area to the containment. The containment of the bloom could be made using circular tongs. Such tongs from the Roman period are depicted on a pottery fragment from Corbridge showing a blacksmith. On the tombstone of a Roman blacksmith found at Dringhouse York (now in The Museum of York) Manning (1976). A similar pair of tongs can be seen. Manning (1985a) Plates 2,3 and 4 show tongs with circular jaws. Tongs are often put to more than one use and could just as easily be used for bloomsmithing. The circular area of the tongs could well have been used to contain bloom iron while it was forged into a compact mass.

Figure 3.26 shows a pair of tongs in the top right of the picture. The jaws of the tongs form a complete circle which could be used for containing bloom iron. If a small piece of bloom was placed inside this ring and a block of iron slightly less than the inside diameter of the ring was placed on top of it, when the top block was struck with a sledge hammer the bloom iron would fuse, because the iron would have no degree of freedom other than compression. Figure 3.27 shows a schematic view through a pair of tongs holding a bloom. (View on the plane A-A).

Finally, there is always the question of whether the original practitioners of bloomsmithing experienced the same problems as present-day experimenters. A bloom from Stadisches Museum, Flensburg, Germany, and is thought to date from the Middle Ages (Röschmann, 1963). It shows clearly, that similar difficulties were encountered in the past, as the condition of this piece of bloom iron mirrors the conditions encountered in my experiments, in that it seems to have fractured in the same manner and the same places as the bloom iron used in my experiments.

The tuyère blocking

The melting clay, either from the floor of the hearth or the tuyère itself, combined with slag, cause the tuyère to block up. The mouth of the tuyère was positioned in the centre of the fuel, and was then in contact with the hottest part of the fire. The various experiments with tuyères have failed to overcome the problem. The blocking of the tuyères was a frequent occurrence in experiments. If this happened in Roman practice then we might expect to find large numbers of blocked tuyères on sites associated with iron-making. The scarcity of tuyères is exemplified in the recent excavations at Woolaston, Gloucestershire, where only a single tuyère was found at a site at which bloomsmithing as well as smelting appear to have occurred (Fulford and Allen 1992). One possible explanation for the low number of tuyères found in connection with bloomsmithing could be because they were never there in the first place. The introduction of air into a blacksmith's forge can be accomplished by means other than a fixed tuyère. Figure 3.26 shows a Roman relief sculpture in the Aquileia Museum, depicting blacksmiths at work. The figure on the left is holding a screen with a Y-shaped protrusion, which he is pointing into a small box-like structure with a steeply pitched roof. The Y-shaped protrusion is a tuyère which, judging by its configuration, is fed from two bellows, thereby giving continual draught. The small box-like structure is a forge hearth, the roof of which will reflect heat downwards, as in a reverberatory furnace. The tuyère is not in contact with the inside of the hearth, and the screen is on legs, so that the whole pumping arrangement can be moved away when necessary. The draught for the forge hearth is clearly from above.

The artifacts identified as tuyères in the archaeological record are made from clay. Use of clay is only necessary if the tuyère is in direct contact with the fire; other materials can be used if the blast is from above. Figure 3.28 shows detail from a fresco in the House of the Vettii, Pompeii, (Etienne 1992, 71). The air blast is being supplied from a tube that the cherub blacksmith has in his mouth. Once again the tuyère is not in direct contact with the fire. If forging is carried out in the box-like structure, then once the inside of the structure is at forging temperature, a modest amount of air blast will be enough to keep the temperature high.

Figure 3.29 shows two cutlers at work. There is an arch between them, which is presumably their forge. It appears to be enclosed and so is more like a furnace. The central figure in Figure 3.30 is standing next to a structure that is presumably man height and encloses a fire. This again is more like a furnace than an open forge.

Summary

When smithing, the volume of air is more important than the pressure. Indeed, experiments by the author have shown that when charcoal is the fuel, high pressure is a definite disadvantage, because the fuel does not have enough weight of its own to prevent it from blowing away if the air pressure is great.

There is positive evidence to show that some billets from the Roman period were made by welding smaller pieces of iron, probably bloom iron, together. Pictorial evidence shows that tongs existed which could have been used to enclose small pieces of bloom iron for forging. The same type of pictorial evidence shows that forging was done in box-like structures, and that air was introduced into the fire from outside. In some cases, the tuyère was not in direct contact with the fire. This has been confirmed by current work by the author (Sim, forthcoming). Together with the evidence of the lack of discarded smithing tuyères from Roman iron-making sites, and the enormous amount of time needed for iron making using the current model, a different model for bloomsmithing is needs to be considered.

Design of Bloomsmithing Hearth No 4

A tapering tuyère was made from 0.8 mm mild steel sheet, with an exit hole of 30 mm diameter. The air blast was directed into the fire at an angle of 50° to the horizontal. With the mouth of the tuyère 100 mm away from the edge of the charcoal, the quantity of cold air passing through the tuyère was sufficient to keep it cool. During experiments the air ran continually for 3 hour 13 minutes, and the tuyère showed no sign of overheating. During smithing operations, it is important that the heat is concentrated in a small area. As stated above, when using charcoal as a fuel for forging, the volume of air is more important than the pressure. A powerful draught is a disadvantage, because it causes the charcoal to blow out of the path of the incoming air. In order to help concentrate the heat, a number of stones were put around the edge of the sand bowl, and a slab of stone was rested on them to act as a roof. This tended to have the effect of a reverberatory furnace and kept the inside at a constant temperature during operation. During both bloomsmithing and welding, the iron was clearly visible in the fire. Moving the iron within the fire presented no difficulties, as there was no adhesion of the charcoal to the iron. Having one side of the fire open made it very easy to rake the iron out for inspection.

Discussion

A possible alternative model for bloomsmithing is suggested. If the bloom is removed from the furnace as soon as smelting is complete, it can be broken up with hammers. The individual pieces can then be forged into cake-like shapes using circular tongs and a hammer. The individual iron cakes thus produced are finally welded together to form a large billet. Any small iron particles too small to weld are returned to the furnace for re-smelting. The air is directed into the forge hearth from above and the tuyère does not come into direct contact with the fire. With this system it is possible to produce 1 kg of bar iron from bloom iron in 2 hours. This is the output of one man working alone, with no mechanical assistance. With one assistant to stoke the fire and watch the iron it would be possible to weld two billets at once and the production time could be reduced by a further 30%. A tuyère which is not in direct contact with the fire will not be blocked by either melting clay or slag. The experiments have shown that a tuyère made of very thin steel will be kept cool by the volume of cold air blowing through it.

If the air is directed down at an angle and the floor of the hearth is sand, then the difficulties of melting clay are overcome. If this is the method used, little or no evidence will exist in the archaeological record. This presents a much more viable time scale than the other models.

3.7 New model for bloomsmithing

The model I propose for bloomsmithing is: The bloom is extracted hot from the furnace. It is broken into small pieces. The small pieces are consolidated in round tongs. The consolidated pieces are welded into a billet. The billet is smithed into bars of various cross sections. These bars are forged into artifacts.

3.8 Experiments using a different approach

Blooms are very delicate when hot and large amounts fall off during forging. It does not seem unreasonable to exploit this effect and turn it to advantage. Tylecote (1976, 41) has suggested blooms were cut up into small pieces for forging. During previous experiments, it was found that a bloom will break into small pieces very easily when heated to red heat (850°C). It was observed in these experiments that the individual pieces were quite well consolidated. This is to be expected as a bloom will break at its weakest points. Most of the pieces were within the range 90 g to 150 g.

Bloom Experiment 10

Considerable insight into the behaviour of bloom iron has been gained from Bloom Experiments 1 to 9. The bloom falls apart when hot and it breaks at its weakest points. This will be where the slag inclusions are large, or where iron has not fused. It seems that this breaking is inevitable, so by making use of it and encouraging it to happen, large slag inclusions will be removed. The slag will be in a semi-liquid form and when the bloom breaks some of the slag will fall off the surfaces, leaving only a small amount sticking to the surface. This is a change in attitude for the smith: much less care has to be taken when smithing blooms and so time is reduced.

Equipment: Bloom hearth No. 4; Fuel charcoal; Block anvil; 5 kg sledge hammer; Axe; Tongs.

The bloom was heated to bright red heat and cut using an axe and the sledge hammer. Very little force was necessary to cause the bloom to break. It broke into twelve pieces of mostly iron. The weights are recorded in table 3.14 below.

Results

Sample Number	weight (g)
1	171
2	155
3	134
4	97
5	135
6	169
7	73
8	113
9	26
10	23
11	42
12	235
Total	1373

Table 3.13 Summery of bloom yield from experiments 1-12

Bloom iron starting weight: 2230 g
Total weight of samples 1 to 12: 1373g.
2230g - 1373g = 875g,
38.43% loss.

In all three experiments were carried out using this method and it became possible to produce 200g of iron in 7 minutes 14 seconds, with a loss of 11% of the material. Previous work by the author, (see section 3.5) and others (Wynne and Tylecote,1958; Cleere, 1971; Crew,1991) have suggested a production time of 4 hours for 200 g and a loss of 75% of original material.

Forge hearth

The design of the bloomsmithing hearth presented many difficulties with tuyère blockage etc. Forging was conducted using the most basic form of forge hearth - a small sand pit, with three sides, the other side open. The air blast was directed from above, through a metal pipe. As it was not in direct contact with the fire there was no need for a clay tuyère. It worked very well and produced more than enough heat to carry out forging.

Bloomsmithing Experiment 11

Objective: To consolidate bloom iron by forging the bloom when it is held inside a pair of round mouthed tongs.

Equipment: Bloom Hearth No. 4; Fuel, charcoal; Bloom iron starting weight 460 g; Block anvil; 5 kg sledge hammer; Axe; Tongs.

A bloom was heated and with very little force was broken into four pieces with tongs. These were all placed into a forge hearth and heated to yellow heat (1000°C). Each piece was extracted and held in tongs which allowed little sideways movement (Fig. 3.27).

Iron weighing 409 grammes was made in 17 minutes 14 seconds, with a weight loss of 11.09%. As before it was easy to see the iron in the fire, there was no adhesion of charcoal and using a rake, it was easy to find the iron in the fire. Having one side of the fire open made it very easy to rake the iron out for inspection and, if the temperature was not high enough, it could be put back with ease.

Bloomsmithing Experiment 12

Objective: To cut up a large bloom with an axe and then consolidate bloom iron by forging the bloom when it is held inside a pair of round-mouthed tongs.

Equipment: Bloom hearth No. 4. Fuel: charcoal. Bloom iron starting weight 1.38 kg. Block anvil. 5 kg sledge hammer. Axe, Tongs.

The bloom was heated for 25 minutes 31 seconds, then taken out of the fire and cut into five big pieces with an axe and hammered at its natural breaking points. The end result was five big pieces and six small pieces with total weight of 1096g. The five big pieces were separated then smithed. The starting weight of five big pieces was 833.0g and after smithing the finished weight was 753.0g. The weight loss was 9.60%

Observations

The tuyère will not burn if it is 50 mm away from the fire; the air passing through it cools it down. If ash is left in the bottom of the hearth it will act as an insulator and the heat will be reflected back up. After the first experiment ash was left in the bottom of the fire. The bloom was extracted hot, and immediately broken into pieces. These were smithed near to the smelting site so that any residue could go back into the fire. The bloom was heated to bright red heat (900°C) and then cut with an axe. The initial blow with the hammer was heavy but then the bloom could be prised apart using the axe head as a lever. It took 3 minutes to cut the bloom into smaller pieces.

The experiments have established that breaking or cutting the bloom into small pieces is a simple and quick matter. It would be more efficient to make use of the heat already present in the bloom to start refining, so the next experiments were conducted breaking the bloom and smithing it in tongs without allowing it to cool.

Bloomsmithing Experiment 13

Objective: To cut up a large bloom of iron with an axe and then consolidate bloom iron by forging the bloom when it is held inside a pair of round-mouthed tongs.

Equipment: Bloom hearth No. 4. Fuel: charcoal. Bloom iron starting weight 3.05 kg. Block anvil. 5 kg sledge hammer. Axe, Tongs.

A bloom with a weight of 3.05 kg was cut up into nine pieces. These were forged in tongs. 1804 g of iron were produced in 1 hour 14 minutes. 375g in smaller pieces were not used.

Observations

When all pieces were in the fire at the same time, it became very confusing and some pieces got lost. It was better to keep them in the fire one at a time, and put the others round the edge to keep hot. The task of controlling the fire and keeping the pieces of bloom hot would be speeded up considerably with the help of an assistant.

The general principle had now been established that blooms could be cut into smaller pieces and forged into well-consolidated iron inside tongs. The next step was to weld the small pieces into a billet. This was done in bloom-welding experiments.

Bloom-Welding Experiment 1

Objective: To determine the time taken to weld small pieces of bloom iron into a billet.

The smithed samples from Bloom Experiments 10 and 11 were used.

A piece of 24 mm square wrought iron bar from Ironbridge was forged to form a plattern to support the pieces that were to be welded. It took 30 minutes 31 seconds to forge to a size of 45 mm x 10 mm x 80 mm. The end was bent to act as a clamp (see Fig. 3.31).

The sequence of operations for Bloom-Welding Experiment 1 are given in table 3.14 below.

Heat No.	Cumulative time (hour : mins)	Tools	Comments
1	0 : 07	Hammer	Weld successful
2	0 : 26	Rake & shovel	Removed clinker
3	0 : 43	Hammer	Weld successful
4	0 : 55	Hammer	Weld successful
5	1 : 10	Hammer	Weld successful
6	1 : 13	Hammer	Weld successful
7	1 : 25	Rake & shovel	Removed P.C.B.
8	1 : 29	Hammer	Weld successful
9	1 : 45	Hammer	Weld successful

Table 3. 14 Sequence and duration of operations in Bloom-welding experiment 1

Total combined weight before welding 1033g
Weight after welding 405g
Loss 60.79% due to welding.
Time 1 hour 45 minutes

Bloom-Welding Experiment 2

Objective: To determine the time taken to weld small pieces of bloom iron into a billet.

A length of wrought iron was flattened at the end to act as a plattern on which to put the pieces of bloom. The end was turned up to act as a clamp to hold the piece of bloom in place. If the bar was placed in the fire with care and, when withdrawn from the fire, it was held down with the end of the rake, the upturn at the end was unnecessary. As the thickness of the billet increased it became easier to weld on the pieces of bloom. This was because the bloom was heating from the bar as well as the fire; indeed, it almost felt like slapping pieces of clay together. There were a lot of sparks when hammering. Leaving small pieces of charcoal on the surface of the bloom seemed to have a cleaning effect on it.

Results

919 g of bloom iron were welded into a billet with a weight of 411 g in 59 minutes.
Loss 55.28% due to welding.

Bloom-Welding Experiment 3

Objective: To determine the time taken to weld small pieces of bloom iron into a billet.

The sequence of operations for Bloom–Welding Experiment 3 are given below in table 3.15.

Heat No.	Cumulative time (min : sec)	Tools	Comments
1	6 : 28	Hammer. Block No.1 (16-10-92)	Weld successful
2	10 : 12	Hammer. Block No 2 (16-10-92)	Weld successful
3	17 : 40	Hammer. Block No. 3 (16-10-92)	Weld successful
4	28 : 31	Hammer. Block No 4 (16-10-92)	Weld successful
5	34 : 28	Hammer. Block No. 5 (16-10-92)	Weld partly successful
6	37 : 12	Hammer. Block No. 5 (16-10-92)	Weld successful
7	43 : 36	Hammer. Block No. 1 (12-10-92)	Weld successful
8	49 : 28	Hammer. Block No. 1 (10-10-92)	Weld successful
9	58 : 26	Hammer. Block No. 2 (10-10-92)	Weld successful

Table 3.15 Sequence and duration of operations in Bloom-Welding experiment 3

Blocks from experiments conducted on 16-10-92 were numbered 1 to 5. Their total weight was 837 g. Blocks from experiments conducted on 10-10-92 were numbered 1 and 2 their total weight was 200 g. Blocks from experiments conducted on 12-9-92 were numbered 1 and 2 their total weight was 156 g.

Results

Starting weight 1193 g
Finishing weight 625 g
Loss due to welding 47.61%

Observations

As the blocks heated up, their surface became sticky. This was because the slag melted and become a treacle-like substance. This stickiness helped to hold the block on the bar when it was in the fire. Welding time was very short, and three hammer blows were usually enough to cause welding. Large amounts of liquid slag came out at high speed causing burns to the operator's hands. The surface of the bar was scraped clean after welding. Welding became easier as the billet got larger. The heat from the billet heated the blocks up, in addition to the heat from the fire, and this reduced the welding time. With the help of an assistant, it would be possible to work two bars at a time.

A section was taken through a piece of iron produced in Bloom-Welding Experiment 2. Figure 3.32 shows that the quality was very high at the bottom of the bloom and decreased towards the top of the bloom. This is because the material at the bottom has had more working and so had more slag expelled than the iron at the top which had only been welded once. Figure 3.33 shows a section of the bloom welded in Bloom-Welding Experiment 3.

3.9 Comparison of results

EXP No.	Initial bloom weight (g)	Finished bloom weight (g)	Material lost (%)	Forging time (hrs : mins)
EXP 1	312	86	72.4	1 : 31
EXP 2	575	143	75.1	1 : 50
EXP 3	203	55	72.9	0 : 52
EXP 4	249	61	75.5	1 : 15
EXP 5	671	170	74.7	3 : 03
EXP 6	No workable iron produced			
EXP 7	No workable iron produced			
EXP 8	No workable iron produced, only iron oxide			
EXP 9	1385	502	63.8	0 : 52
EXP 10	2230	1375	38.43	1 : 37

Table 3.16 Results from Bloom-welding Experiments

EXP No.	Initial bloom weight (g)	Finished bloom weight (g)	Material lost %	Forging time (mins)
EXP 11a	85	74	12.94	4.0
EXP 11b	82	75	8.54	4.0
EXP 11c	135	129	4.44	5.0
EXP 11d	158	131	17.09	4.0

Table 3.17 Results of experiments conducted in round tongs.

EXP. No.	Initial weight of bloom (g)	Finished weight of bloom (g)	Material lost (%)	Forging time (min)
EXP 12	99	89	10.10	5.0
	118	102	13.56	6.0
	182	168	7.70	6.0
	185	163	11.89	5.0
	249	231	7.23	7.0
EXP 13	The weights of the nine individual pieces were not recorded.			
	2380	2179	8.45	31.0
EXP 14	667	521	21.89	9.0
	346	303	12.42	8.0
	165	160	3.03	3.0
	134	127	5.24	4.0
	76	59	22.37	4.0

Table 3.18 Results of smithing a single bloom in tongs.

EXP. No.	Initial weight of bloom. (g)	Finished weight of bloom. (g)	Material lost (%)	Forging time. (h : m)
BWEXP. 1	1033	405	60.79	1 : 45
BWEXP. 2	919	411	55.23	0 : 59
BWEXP. 3	1193	625	47.61	0 :34

Table 3.19 Results of bloom smithing in tongs

Using the traditional method of bloomsmithing, the best results were 0.0305 minutes to produce one gram of iron, against 1.04 minutes per gram using the traditionally accepted methods of bloomsmithing. With the alternative system, one kilogram of iron can be produced in 30.5 minutes. This will give a production time of 508.33 hours to produce one tonne of iron.

Considering again Aiano's (1975) proposed production figure of 2 250 tonnes of iron/year in Britain. Using the alternative model, 2 250 tonnes of iron per year can be produced in 1 143 750 hours as opposed to 39 000 000 hours using the traditional method. This is a reduction of factor 34.1.

From these experiments, a much more realistic figure for the time to process bloom iron has emerged. The figure is still very high, but it was arrived at through experiments conducted by one man working alone, using only hand methods. It is a matter of pure speculation how much mechanical devices might have improved production time. On present evidence, such devices are not thought to have existed in the Roman period but, given the amount of iron produced and the time period over which it was produced, the possible existence of mechanical devices is a matter for future research.

The training of a team of men to process bloom iron would also have a significant effect on production time, but how much more time would be saved, can only be determined by further experimentation.

3.10 Discussion

The processing of iron ore into iron artifacts can be considered as three separate operations: smelting; bloomsmithing; and blacksmithing. All three are specialised processes and require specialist knowledge. All three require not only special operatives but special equipment.

In the case of bloomsmithing, the various hearths considered in section 3.2 show that special furnaces were constructed for use in iron making, but their use for bloomsmithing is by no means proved. The experiments using reconstructions of some of these furnaces have shown that, at least under experimental conditions, these furnaces proved to be ineffective, because of the continual blocking of the tuyère. No production process can be effective if the equipment fails after only forty minutes and requires several hours to repair. Such a system would not be able to run for long enough to meet demands, and the demand for iron in the Roman period is known to be high. Tylecote (1976) notes that the amounts of slag found on Roman sites is measured in hundreds of tonnes. The annual output for Britain has been estimated at 2 250 tonnes (Aiano 1975), and for the Empire 82 500 tonnes (Healy 1978, 196). Further, the original practitioners would not work with this sort of continual failure, and would have found ways of rectifying the problems.

It would therefore appear, that whatever these furnaces were used for, the way in which they were used is of greater significance, (Sim, forthcoming). The critical factor is the use of a built-in tuyère to direct air into the fire. With this system, the tuyère is blocked by a combination of melting clay and slag liberated from the bloom. Clay and slag combine and cause a blockage of the tuyère which adheres so strongly to the wall of the furnace that the furnace is often damaged beyond repair when the blockage is removed. It is unrealistic to suppose that this was the operating system.

The experiments with replaceable tuyères (see section 3.4) show that the same blocking takes place over the same duration of time, but that a blocked tuyère can be replaced very quickly. These tuyères can be produced very quickly, and the smith could have easily made and fired them himself. If this system was used, bloomsmithing would use as many as sixteen tuyères a day, and one would therefore expect to find vast numbers of tuyère remains on archaeological sites. Even if the body of the tuyère has disintegrated, the mouth would remain because it had vitrified. Vitrified material remains better preserved in the archaeological record. As tuyère finds are comparatively rare, the assumption must be that this was not the system in use.

Pictorial evidence from Aquileia and Pompeii (see Figs. 3.26 and 3.28) show a possible alternative and demonstrate that forging was possible with a tuyère not in direct contact with the forge fuel. In one instance (Fig. 3.26), the blast is supplied by a pair of bellows,(held by the figure on the left of the picture). In the other (Fig. 3.28), the air is supplied from a pipe that is blown by mouth (it is still common practice among modern silversmiths to use spirit lamps and a blowpipe for silver soldering). In both instances, the hearth was enclosed and had to be well insulated, so that once the fire was hot, it would require minimal air blast to keep the temperature at the necessary level. An enclosed hearth with air supplied by a tuyère which is not in direct contact with the fuel will overcome all the difficulties that the accepted model experiences.

The experiments in bloomsmithing showed that several different approaches failed to overcome the basic problem of the bloom falling apart before it could be consolidated enough to be smithed with sufficient force to expel the slag and weld the iron together. The time to arrive at workable iron was enormous and, following the current model, a time to ***smelt and smith*** enough iron for the annual needs of Britain would have been **80 000 000 hours**, a figure which seems unrealistic.

The fundamental problem is to provide the means of administering the force with a surface area greater than that of the bloom and restricting the degrees of freedom of the bloom during forging. The experiments using tongs reduced the bloom smithing time by a factor of 34, which is a more realistic figure. When this method was combined with an enclosed hearth and a tuyère which was not in direct contact with the fuel, it becomes a much more efficient system that is likely to have been able to meet the demands of the period.

A consolidated billet of iron can be used almost as delivered if it is to be used for large iron items such as hammers (see chapter 4.4), but when smaller artifacts are needed, the billet it has to be forged down to a size which is easy to handle and can be heated in a fairly short time. Possibly the smithing of blooms down into bar was carried out by teams of semi-skilled strikers led by a skilled blacksmith.

The work so far has shown that a more realistic model for iron processing after smelting (bloomsmithing) is best conducted using an enclosed hearth with air supplied by tuyères that are not in direct contact with the fuel. Blooms were broken into small pieces and consolidated in tongs. It is presumed that billets were built up from small pieces welded together and these billets were then delivered to the blacksmith for forging and manufacturing of artifacts.

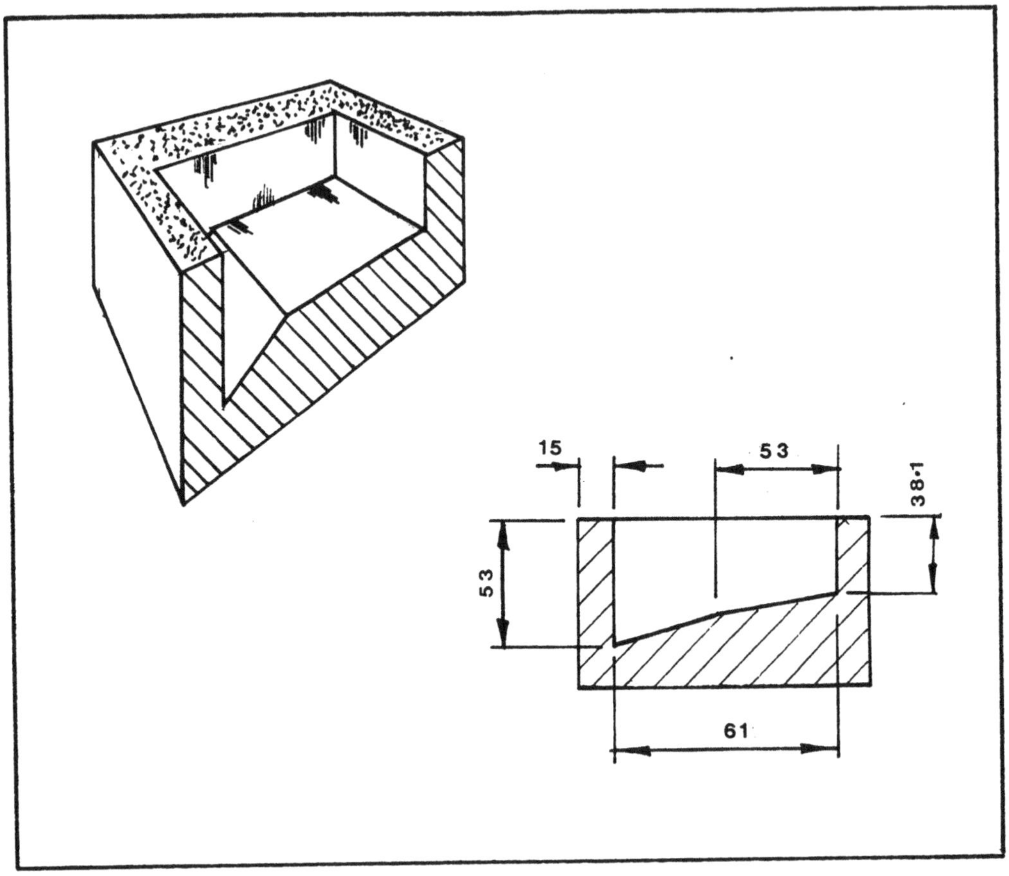

Figure 3.1 Profile of the furnace found at Woolsthorpe.(drawing not to scale)

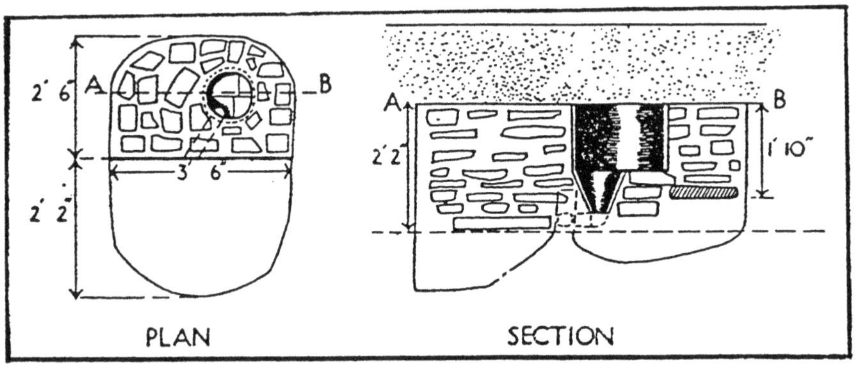

Figure 3.2 Cross section of a furnace from Wilderspool. (drawing not to scale)

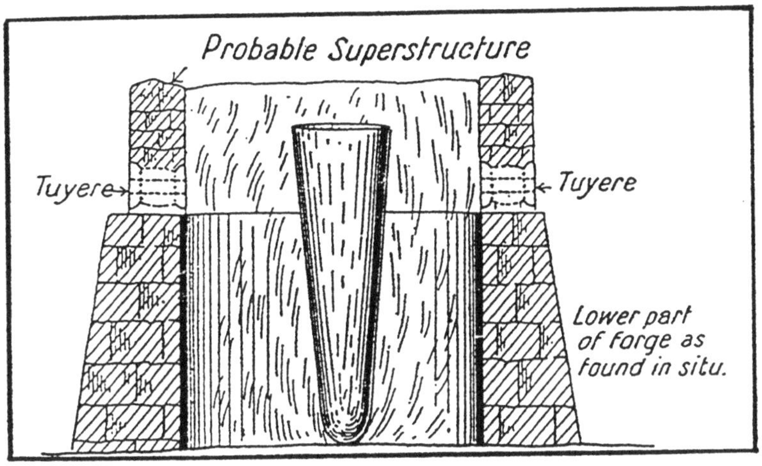

Figure 3.3 Furnace for welding bath house beams. (Singer *et al* 1954, 51)

Figure 3.4 A black figure painting on a 6th Century BC vase, showing the forge of Hephaistos. (British Museum)

Figure 3.5 Bloomsmithing .from Eisengewinning in norddentschen Flachland. By Hauke Jöns. In Alter Bergbau in Deutschland Archaeologie in Deutschland Sonderheft. (1993, 39)

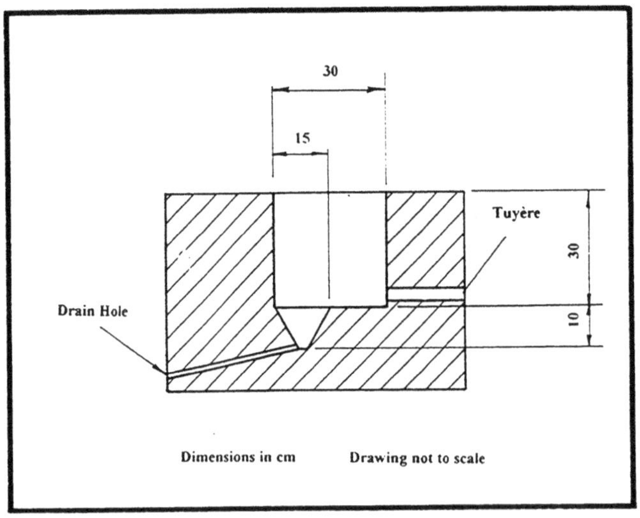

Figure 3.6 Section through Experimental Bloomsmithing Hearth No. 1.

Figure 3.7 Slag build-up around the tuyère of Experimental Bloomsmithing Hearth No. 1.

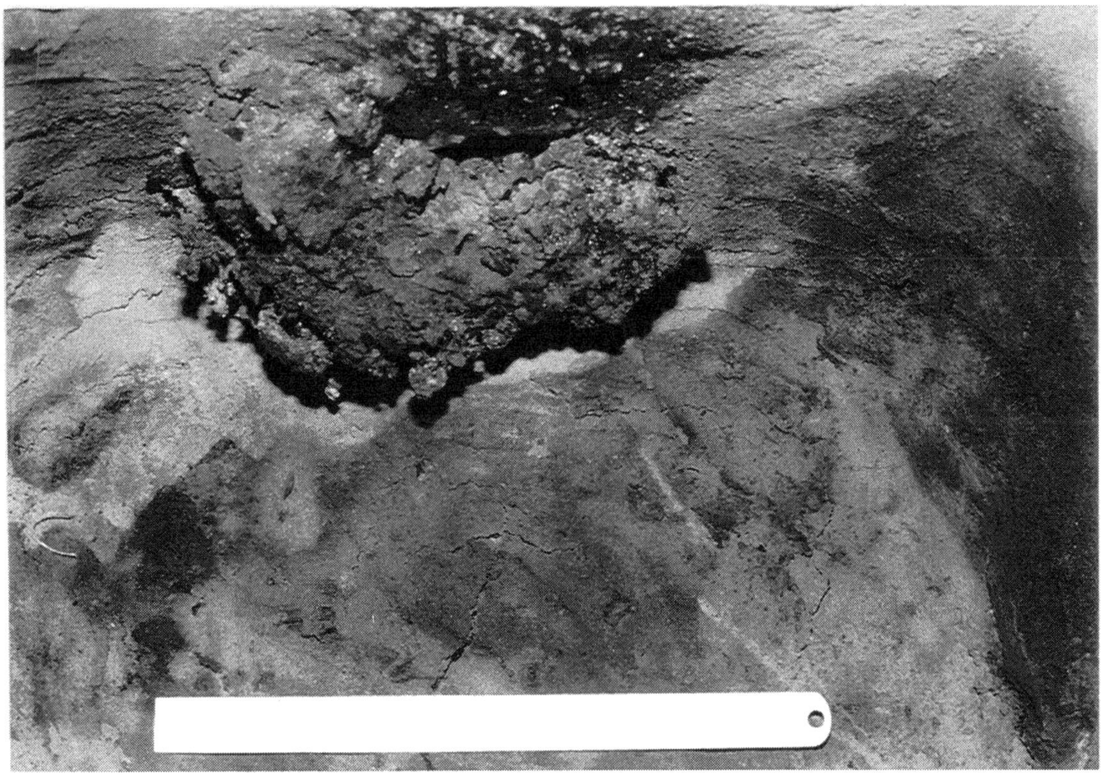

Figure 3.8 Slag build-up around the tuyère of Experimental Bloomsmithing Hearth No. 1.

Figure 3.9 Slag has come away cleanly from the tuyère of Experimental Bloomsmithing Hearth No. 1.

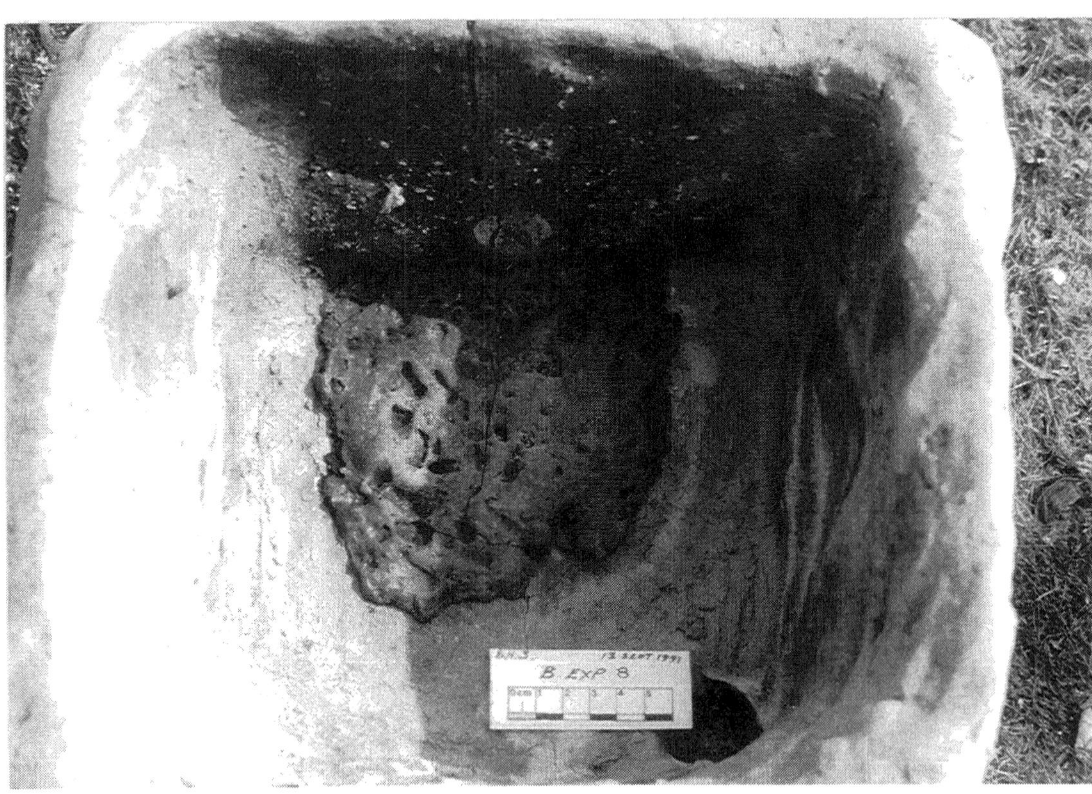

Figure 3.10 Experimental Bloomsmithing Hearth No. 1 after removal of the bloom showing crack.

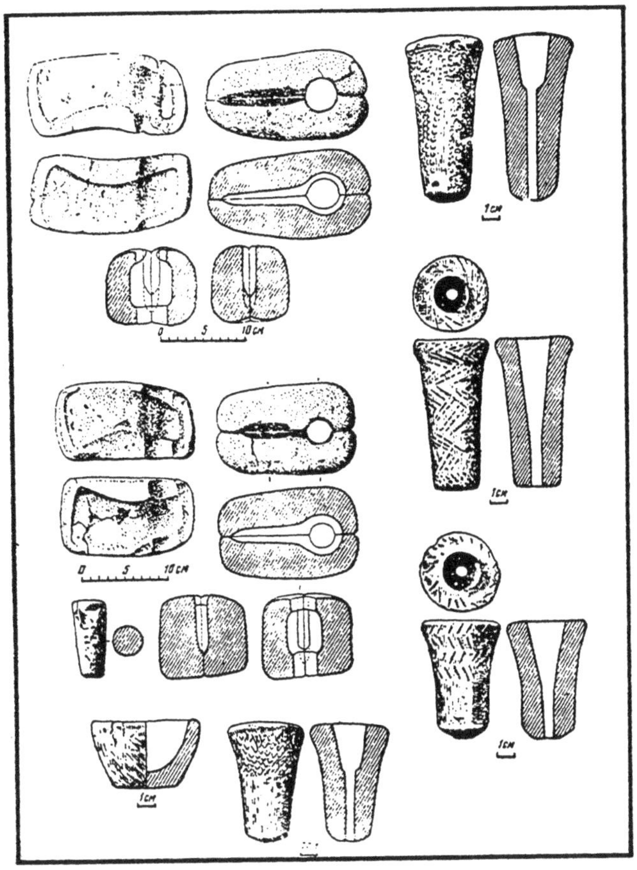

Figure 3.11 Smiths tuyère from Kalinovka, USSR (after Mongait)

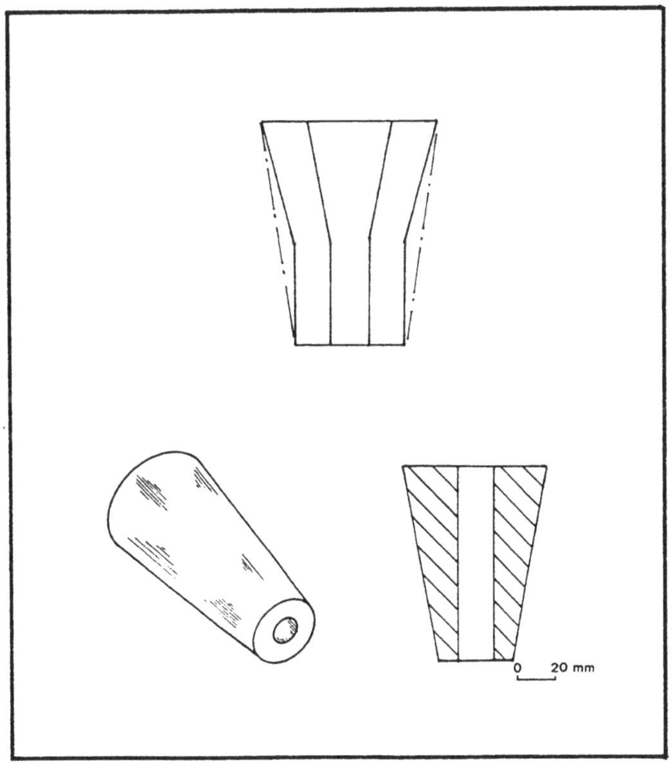

Figure 3.12 Tuyère in its simplest form.

Figure 3.13 Blocked tuyère

Figure 3.14 Section through a piece of 'muck bar'.

Figure 3.15 Section through a piece of five rolled wrought iron.

Figure 3.16 Section through a piece of Roman iron.

Figure 3.17 Polished section of bloom from Experiment No. 1.

Figure 3.18 Polished section of bloom from Experiment No. 2.

Figure 3.19 Polished section of bloom from Experiment No. 3.

Figure 3.20 Section through bloom from Experiment No. 4.

Figure 3.21 Polished section of bloom from Experiment No. 5.

Figure 3.22 Polished section of bloom from Experiment No. 9.

Figure 3.23 Billet from Newstead number FRA 325.(Courtesy of National Museum of Scotland)

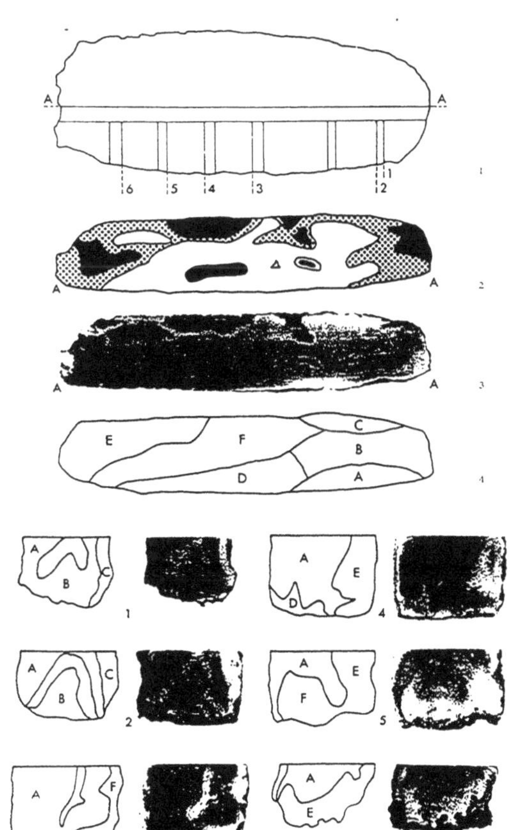

Figure 3.24 Billets from Haithabu, made from small pieces welded together. (after Thomsen *et al* 1971)

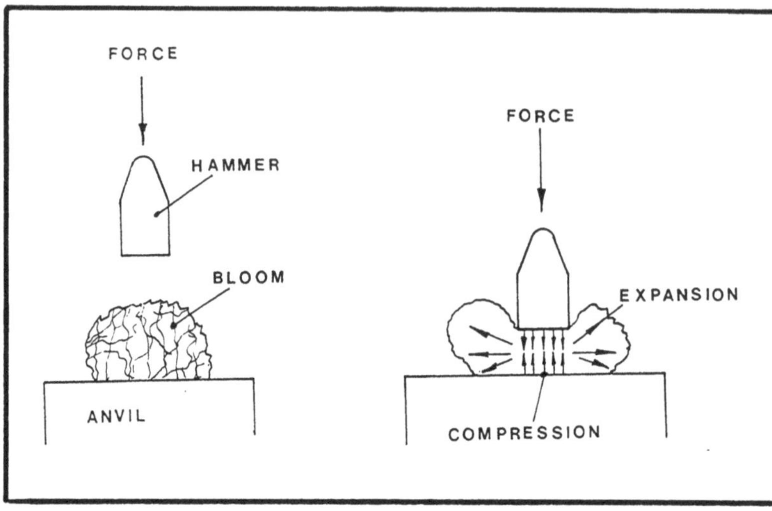

Figure 3.25 Schematic view of bloom iron during hammering

Figure 3.26 Tongs shown in the top right of the picture form a complete circle.

(From a Roman relief sculpture Aquileia Museum.)

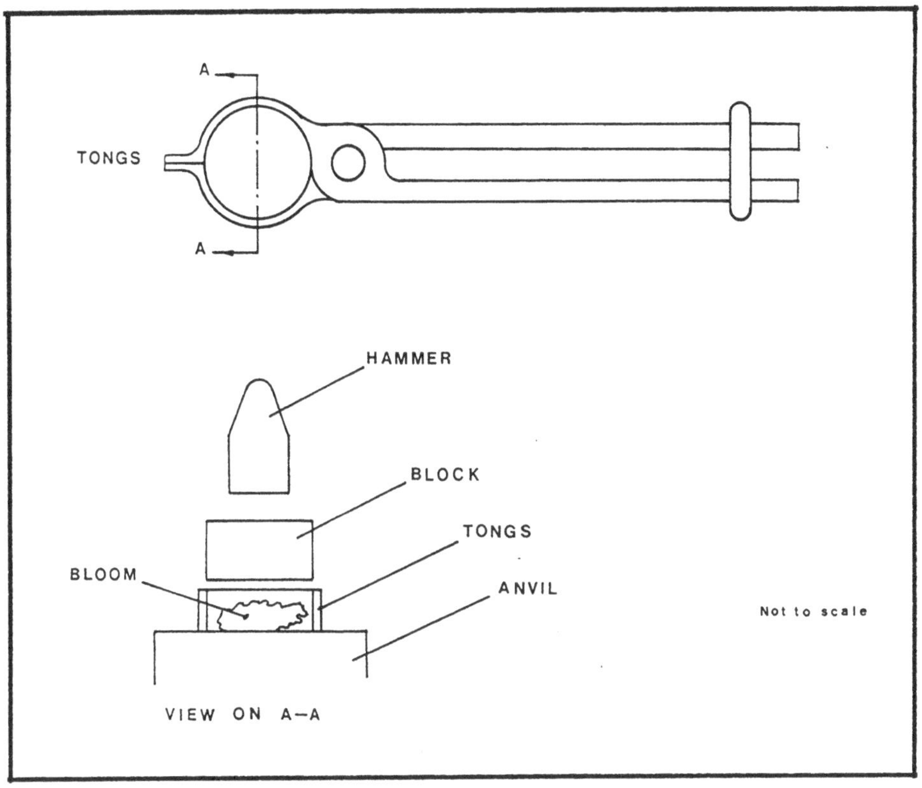

Figure 3.27 Schematic view through a pair of tongs holding a bloom.

Figure 3.28 Details showing the use of a blow-pipe, from a fresco in the House of the Vettii, Pompeii.(Etienne,1992,71)

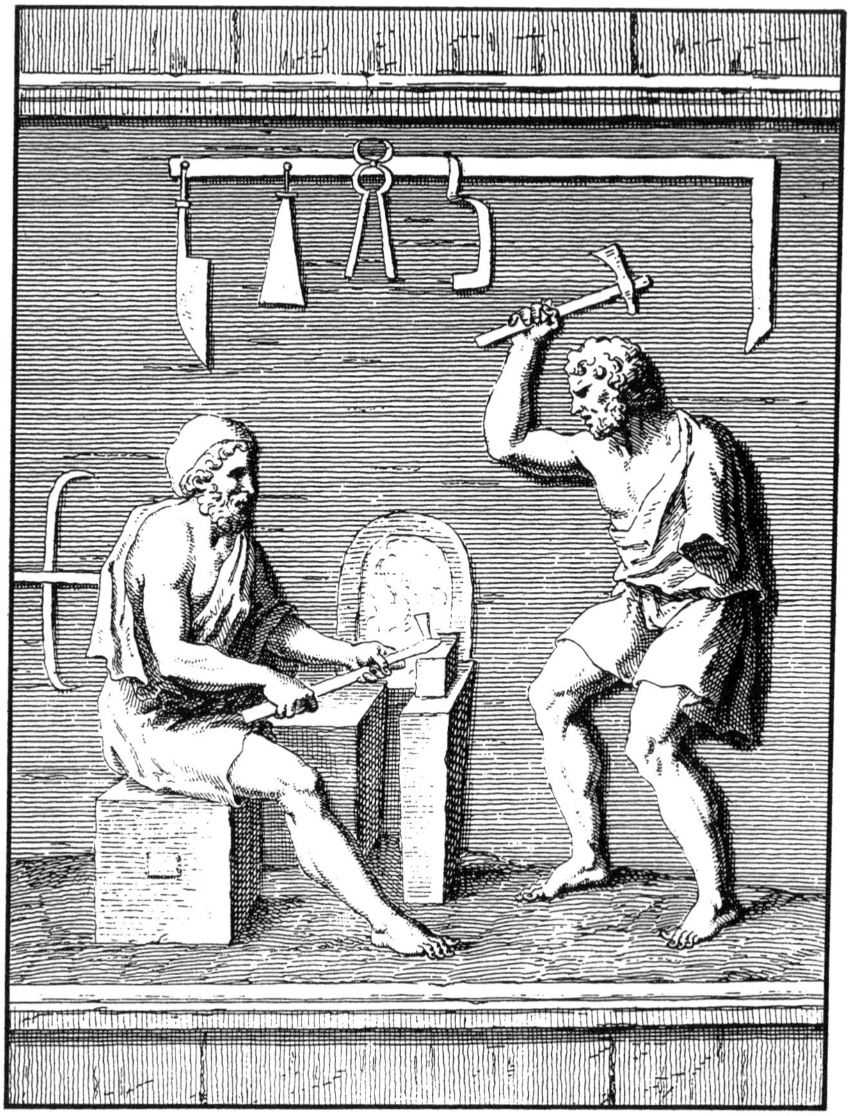

Figure 3.29 Two cutlers at work. (From a mural in relief at the Villa di Montalto, Rome.Entitled Vulcans forge.)

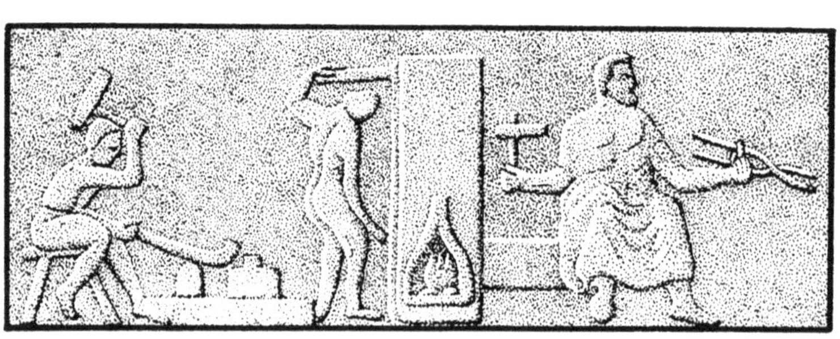

Figure 3.30 Detail from a Greek bowl.(Louvre, Paris)

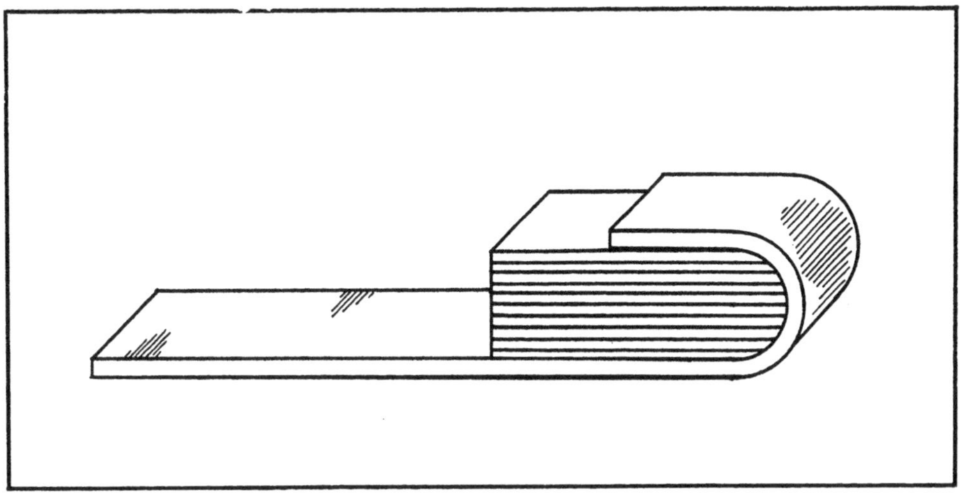

Figure 3.31 Plattern bent to hold small pieces of bloom iron.

Figure 3.32 Section through Bloom Weld 2.

Figure 3.33 Section through Bloom Welding 3.

CHAPTER 4

Roman iron artifacts

'It is with fire that blacksmiths iron subdue
Unto fair form, the image of their thought:
Nor without fire hath any artist wrought
Gold to its utmost purity of hue.
Nay, nor the unmatched phoenix lives anew,
Unless she burn.'

Michelangelo, *Sonnet 59*

4.1 Introduction

This chapter will detail the experiments which were conducted to produce copies of Roman iron artifacts. It will consider the time taken to refine billets of iron into workable iron and the time to produce specific iron items.

Aside from the cost of raw materials used in the production of iron items, the cost of man power is a major contributor to the value of individual iron artifacts. Copies of Roman iron artifacts were made with copies of the tools, materials and equipment available during the Roman period and a record was made of the time taken for manufacture and the resources consumed.

There are three basic methods of forming metal into artifacts: casting; metal removal; and, metal reshaping i.e. forging (see glossary). In the Roman period cast iron was not produced deliberately (Tylecote 1976). Metal removal means cutting away material to form a shape. In the Roman period this would have been done by hand and would have be very time-consuming. In antiquity, the principal method of forming iron into artifacts was by forging. This remained true until the development of cast iron in the 14th century. Powerful machine tools in the present era opened the way to new methods of making iron artifacts. Forging is a method of manufacture in which metal is made soft by the application of heat, and is then shaped by the use of suitable force, often supplied by a hammer. The basic principles of forging and the terms used in the craft of the blacksmith are covered in the glossary of this book.

Many of the iron artifacts recovered from the archaeological record are often little more than a mass of iron oxide, the shape of which gives no indication of its previous structure and function. When compared with non-ferrous metals, iron lacks durability which leads it to be regarded as somehow inferior. This may be the reason why some products made from iron are often thought of as being the "cheap end of the range" (Manning 1985a). Indeed, when compared to many items made from non-ferrous metals, iron artifacts are relatively cheap to make, because forging is a faster and less complex system than casting.

Many iron items may have been made by the local blacksmith, and individual small workshops were capable of supplying the needs of both the local community and the smaller towns. A system of this sort will work on a small scale. In the context of the Roman army, iron weapons were probably made on a large scale, and it seems unlikely that small workshops would have been able to meet the demand.

The production of iron items such as weapons and tools which were supplied both to individuals and the Roman army forms the basis for this chapter. A number of fundamental questions are addressed:

(a) What level of technology was available to the blacksmith in the Roman period?

(b) How long did individual items take to make?

(c) Were all iron items produced by skilled craftsmen, or were semi-skilled workmen involved?

(d) How did the level of demand affect the system of production?

To attempt to answer these questions a series of practical experiments was undertaken, manufacturing a series of artifacts in everyday Roman use using only materials, tools and fuel available at the time.

There is evidence from the archaeological record which will assist in these experiments, although much information will be gained from making the items. We know that the principle metal used was wrought iron. Steel was expensive and was used sparingly. Air was supplied by hand-operated bellows.

Manning (1972, 55) argues that many Roman tools are identical to their nineteenth-century successors. We can also say with confidence that many of the tools of the Roman blacksmiths are identical in appearance to those of the 20th century blacksmith. Indeed, Roman blacksmiths' tongs represent the terminal development of tongs (Olhaver 1939). After a study of the iron from the Fort of Saalburg, other collections at the British Museum, and Reading Museum, I concluded that, with the exception of some types of hammer, (see chapter 4.3) all the tools required for the experiments detailed below were already in my existing kit of blacksmith's tools. The range of hand tools available to the Roman blacksmith was as extensive as that of a modern blacksmith, the exception being the twist drill, which is not found in the Roman period. Although the Roman blacksmiths' tools are still in existence, the working practices they followed are not so readily available. For example there is no indication as to how many pieces of iron a blacksmith may have in the fire at any one time. The modern practice is for the smith to have one piece he is working on while a second is heating up. When assistants are available, they perform tasks such as bringing in fresh fuel and stoking the fire. Since there are no surviving written records, or manuals, describing the working practices of Roman iron makers or blacksmiths, we must make assumptions to establish what their working practices may have been. Before experiments to reproduce copies of Roman iron artifacts could be started, decisions had to be made regarding the protocol to be used in all experiments. It is safe to assume that the basic principles governing the forging of iron (hot iron was hit with a hammer) were exactly the same then as now. I have assumed that working practices required that, items be made with the minimum wastage of metal, whilst using the minimum quantities of fuel and taking the minimum time. The day-to-day operations of forge management were are also assumed to be the same system in operation today, based on the blacksmith avoiding injury to himself and working as efficiently as possible.

Experiments to determine a suitable type of forge hearth have already been discussed earlier in Chapter 3. Much of the fuel used was charcoal. Although in some instances coal is known to have been used (Webster 1955), all the following experiments were conducted using charcoal.

The standard of workmanship is difficult to gauge. For prestigious items such as dress weapons, the standard must be as close to perfection as possible, but for disposable items, such as regular-issue weapons, the standard must be good enough for it to perform its function efficiently and correctly. It is possible that individual buyers set their own standards for their suppliers and this would mean that standards differed from one area to another, and from one time to another, as the buyers changed. Accordingly, I set my own standards by asking what would be acceptable to me if I were the buyer.

Costs

When considering the cost involved in the manufacture of the artifacts made in the experiments I considered the two major factors of labour and material costs. The factor of profit has not been considered because there is too little information available from antiquity to make any firm assumptions regarding assessment of profit.

Labour force

The Roman iron industry appears to have two major sources of labour available - free men (including legionaries), and slaves. Irrespective of the worker's status, some factors are common to everyone who was involved in ironworking. Most aspects of iron making required skilled and semi-skilled workers to perform the tasks. This involved training workers who would be unproductive during training and would take time to practise their skills until they reached a standard when they gave a return for the investment in their training. Semi-skilled workers could be trained in a reasonably short time, but a skilled iron worker takes a long time to train. The necessary skills are very demanding both mentally and physically, and cannot be taught to everyone. Most of the workforce were likely to be skilled workers.

Free men would demand a wage of some sort, and even a low wage contributed to the cost of making iron. The position of slave labour is not as clear-cut as it first seems. Even if all the workforce were slaves, they still had to be fed, clothed and sheltered, all of which involved expenditure. Thus the running costs for a slave could be quite high and even though slaves may not have been given an actual wage, there was still a cost involved. To what extent the army was able to meet its own demands is unclear. The legionary was already being paid for being a soldier - perhaps the cost of labour for in-house iron items was slight. The running costs for a slave could be quite high and this meant that even though slaves may not have been given an actual wage, there was still a cost involved. All iron artifacts produced in the Roman period had to be paid for, and the payment was either in cash or in kind.

The composition of the labour force is not clear when dealing with the Roman army. It is possible that some iron items were made by *immunes* (see glossary). They might also have carried out repair and maintenance of weapons, as is suggested by the iron items found at Corbridge (Alleson-Jones and Bishop, 1988). When supplying a large standing army, although some weapons are recycled, a large number have to be replaced because of loss in action; theft; breakage; and, loss in camp from stores and supplies. It is impossible to know the size of the workforce in any individual workshop, or how much civilian labour was used, so again assumptions have to be made, but these can be based on recent evidence for the working practices of modern blacksmiths. The minimum viable workforce is a skilled blacksmith assisted by a single striker. As the blacksmith gains experience, production times for making particular artifacts will diminish. After a number of items have been made, the time will level off and become more or less constant.

In some of the experiments, it will be possible to make more than one of the same item, but it will not be possible to make large enough quantities so that through practice, production times will be reduced to the minimum. However, it will be possible to estimate with a great deal of accuracy the expected times for production of most artifacts. This will give the time for the minimum workforce described above to make any specific item, and simple calculations will show the times for a larger workforce.

The experiments started with the time required to forge billets into workable bars and followed with experiments to produce the following artifacts:

(a) a blacksmith's hammer;

(b) a stylus;

(c) a series of Roman nails;

(d) a series of disposable weapons including: pilum, balista bolts heads (square and pyramid), flat bladed bolt head, spear head, fire arrow head;

(e) a pattern welded sword; and,

(f) chain mail.

The above list represents a broad cross section of items. The stylus would have been an item found in almost every house hold, whilst the blacksmith's hammer was for a more specialised use. Nails would have been used in great quantities where ever timber was in use, for domestic buildings, in ship building, and by the army for the construction of timber forts. The series of disposable weapons examines another aspect of manufacture: the need to make items which were not designed to be kept, and which would probably have a short service life. At the other end of the spectrum a pattern welded sword and chain mail coat represent artifacts which were required in some quantity, but involved much time in manufacture and would be valuable items.

All of the items to be manufactured would have a common starting point: the bloom of iron.

This chapter goes beyond the bloom - it considers the delivery and forging of a billet to produce workable bars of varying dimensions which would be used to manufacture various artifacts. The following sections consider first bloom refining and then the production of each of the artifacts listed above.

4.2 Forging billets into bars

Delivery of material to the blacksmith

It is assumed that the legionary smith received his iron in the form of billets. Evidence for this comes from auxiliary forts in northern Britain such as Newstead, (Curle 1911, 288 pl. LXV, Fig. 9), and Strageath (Frere and Wilkes 1989, 167, pl. XII) where iron bars attributed to the Flavian period have been found (Fig.3.23). Tables 4.1 and 4.2 give the sizes and weights of these bars. The average weight of these bars is 6.79 kg. The greatest weight above this is 7.4 kg, the lowest 6.0 kg. This variation, combined with the similarity in dimensions, can be construed as a form of moderate standardisation.

Before the production of individual weapons or artifacts, it is necessary to reduce the billet to the appropriate size and shape depending on the form of the finished item. The production of a pilum, for example requires bars to be forged, whereas the production of arrow heads requires iron sheet.

Dimensions (in cm)	Weight
39 x 6 x 6	7.0 kg
34 x 6 x 5	6.0 kg
34 x 6 x 5	6.5 kg

Table 4.1 Sizes and weights of iron ingots from Legionary Fortress at Newstead (Curle, 1911).

Dimensions (in cm)	Weight
30 x 6.5 x 6.5	7.3 kg
36 x 6 x 6 (each end tapers to 4 x 4)	5.7 kg
45 x 6 x 7 (each end tapers to 3 x 3)	6.9 kg
34 x 6 x 6 (each end tapers to 5 x 5)	7.4 kg

Table 4.2 Sizes and weights of iron ingots from Legionary Fortress at Strageath. (Frere and Wilkes 1989)

Experiment 1: Reduction of Billet

This experiment was conducted to determine the time required and the charcoal consumed in forging down a billet of the same size and weight as those recovered from Strageath. A billet of wrought iron weighing 5.8 kg was used. After one hour the billet was only dull red (800°C). The continual re-heating after hammering of such a large piece of iron in a normal blacksmith's forge hearth would take too long to be practical. This problem could be overcome if the billet were smaller. I decided to cut the bar into smaller and more manageable pieces. It took 1 hour 18 minutes to raise the temperature of the billet to a heat where it was possible to cut it with a hot set and sledge hammer. A piece weighing 1.22 kg was cut off. The cutting off took 58 seconds and 45 hammer blows. This produced a piece of iron with dimensions of 55 mm x 28 mm tapering to 45 mm x 30 mm and a length of 136 mm. This was forged to a regular cross-section of 30 mm x 55 mm in a time of 22 minutes 20 seconds. This billet was then forged down to cross-sections of various sizes. Table 4.3 presents the data obtained during this experiment.

Column 1 is the number of times the billet was heated.

Column 2 gives the time to re-heat the billet after hammering.

Column 3 is the time the iron was under the hammer.

Column 4 the cumulative time of heating and hammering.

Column 5 the dimensional changes the billet underwent during heating and hammering.

Use of this table can give a good approximation of the time required to prepare a blank from a billet. For example: to prepare a blank of 20 mm x 20 mm x 250 mm long from a billet 55 mm x 30 mm x 136 mm requires almost three hours (see heat number 15 in table 4.3 179 mins. 04 secs.).

Heat No.	Time to reheat (m : s)	Time under hammer (m : s)	Cumulative time (m : s)	Dimensions (mm)
0	-	-	-	55x28x136*
1	19 : 57	2 : 23	21 : 08	55x30x136
2	6 : 28	2 : 01	30 : 09	42x28x146
3	14 : 50	1 : 45	46 : 04	38x23x150
4	5 : 54	1 : 58	53 : 16	37x34x160
5	8 : 10	2 : 11	63 : 37	35x40x160
6	11 : 41	2 : 17	77 : 35	33x23x170
7	5 : 20	2 : 41	85 : 36	26x32x185
8	7 : 31	2 : 25	95 : 32	30x28x195
9	7 : 53	1 : 52	104 : 37	28x28x200
10	3 : 40	0 : 34	108 : 11	28x28x203
11	6 : 40	3 : 46	118 : 37	26x29x210
12	9 : 37	3 : 50	131 : 24	25x28x216
13	9 : 36	5 : 00	145 : 06	25x25x220
14	13 : 10	5 : 00	163 : 07	22x22x230
15	10 : 52	4 : 42	179 : 04	20x20x250
16	15 : 06	3 : 20	197 : 30	26x15x290
17	10 : 44	4 : 00	212 : 14	15x15x320
18	10 : 14	2 : 40	225 : 18	15x13x330
19	10 : 02	2 : 34	237 : 02	13x13x360
20	9 : 53	2 : 26	249 : 21	11x11x390
21	9 : 27	2 : 41	261 : 29	9x9x500
22	8 : 46	2 : 10	272 : 25	7x7x650
23	9 : 46	2 : 56	284 : 27	6x6x800

*NB the billet forged in this experiment was initially tapered (55 mm x28 mm at one end and 45 mm x28 mm at the other) before being forged to a uniform section 136 mm long.

Table 4.3 Forging of a billet

It is interesting to note that, during the forging process the iron billet lost 19.75% of its original weight, and about 84% of its original volume. This was caused by slag and air exclusion from the material, as well as material loss.

Discussion

As no evidence has so far come to light for the use of mechanical hammers in the Roman period, we must assume that manpower supplied the energy needed to refine the billet. It was common practice until recently for a blacksmith to have at least one striker (see glossary) working for him. There is pictorial evidence to show the use of strikers in antiquity (Healy 1978, 164, 52), and indeed it is not possible for one man alone to perform many of the different operations used in blacksmithing. The number of strikers is open to question. In a German wood-cut (Amman 1989, 70) three strikers are shown working together with a scythe maker. The shape and size of the anvil will determine the number of strikers who can physically fit round it. The Roman anvil and the one used by the scythe maker are similar in size and shape, it is possible that as many as three strikers could have been employed in Roman times to refine blooms. Striking is a very skilled task; but it is skill in a very specific and narrow area, and can be taught quickly. It does not compare with that of a blacksmith, who has to have a wide area of expertise, which takes a considerable time to acquire.

A number of strikers can be trained as a team in a very short space of time - two to three days would be quite sufficient. At the end of this period, they will be able to work efficiently, under the orders and direction of a blacksmith. Therefore it follows that forging billets into specific cross-sections is a skill which can be easily taught. It is possible that when an army was in the field, there were *immunes* who dealt with forging bars to the size necessary for the making of specific weapons. The forging of special sized bars required more skill and would have been the responsibility of the more skilled smiths. During the experiments only one striker was employed, but simple multiplication will give the amount of work done by the numbers of strikers. In the early stages, the length of heating time is considerable, due to the large volume of metal being heated. During heating time the workforce is idle, so it is possible that the team would work with two bars; one being heated, while the other is being forged. In this way manpower is used more efficiently. It is unlikely that any team would work on more than two billets at once. Striking is very hard and tiring work, and the striker, no matter how strong, will soon become exhausted by continual work without rest periods. If the striking team is given a short rest between hammering sequences, they will be able to work for a much longer period without fatigue.

Although it is possible to work the whole billet at once, it was probably not the common practice to do so. The billet as a whole, is heavy and very difficult to manipulate. When the billet is whole it requires a considerable time to bring up to forging heat and although this time is shorter during subsequent heats, the heating still consumes a large quantity of fuel and time. Although there is no direct evidence of billets being cut up, evidence from the 15th century shows that it was common to supply the smith with billets about one third the size of the Roman billet (Lapsley 1899, 510). A combination of ergonomics and common sense show that a billet of 2.1 kg is much easier to heat and handle than one three times the size.

Summary

Iron was probably supplied to the Roman blacksmith in the field in the form of billets. The quality was sufficiently high for the smith to manufacture some artifacts straight from the billet. Billets were smithed down into cross-sections which provided blanks for the subsequent forging. The forging of billets into bars is a semi-skilled operation, and operators could be trained in a short space of time. The billet smithing was carried out by a smith and up to three strikers, and not more than two billets were worked by any one team. The preparation of blanks consumes a large quantity of both fuel and manpower and this is a significant factor when the total production time for an artifact is considered.

4.3 The reproduction of a Roman hammer

Almost every craft that requires the use of hammers has developed special hammers to fulfil particular needs. This was so in antiquity, as can be seen from the various hammers documented by Olhaver (1939), Manning (1976, 1977, 1985a), and Gaitzsch (1980). The ancient blacksmith would have made hammers for other craftsmen as well as for himself.

In order to reproduce artifacts in the same manner as a Roman blacksmith, working conditions and tools had to be as close to the originals as possible. To meet this requirement, replicas of Roman blacksmiths' tools were made, including a hammer.

Figure 3.29 which is a Mural in relief at the Villa di Montalto, Rome, shows two cutlers at work. The figure on the right of the picture is holding a hammer with the shaft protruding above the head. This suggests that he is using a hammer with the eye in the form of a truncated cone. This is a type of hammer used by Roman blacksmiths was the design used for this experiment. A Roman hammer from the Silchester collection (ref. 07510 yc 169) is of this type. With the permission of the curator of Reading Museum, the hammer was examined, measured and weighed. An outline drawing of the hammer can be seen in (Fig. 4.1). This information was used to make a copy of the hammer.

Making the hammer

Equipment: Forge hearth, Anvil, Cross-peen hammer, Bottom swage (round), Hot set, Tongs, Fuel: charcoal.

The starting point was a piece of well-consolidated wrought iron weighing 4 kg.

Operation 1	The billet was forged into a bar with a cross-section slightly larger than the final cross-section of the finished hammer (Fig. 4.2 a).
Operation 2	The peen end of the hammer was forged (Fig. 4.2 b).
Operation 3	The radius on the end was forged (Fig. 4.2 c).
Operation 4	The position of the eye-hole was marked and a small hole was punched through the thickness of the body of the hammer.
Operation 5	The eye-hole was formed using a taper punch (Fig. 4.2 d).
Operation 6	The hammer head was cut from the bar, an allowance of 20 mm was made on the length (Fig. 4.2 e).
Operation 7	The striking end flat was hammered flat (Fig. 4.2 f).
Operation 8	The hammer was heated to dull red and the scale was scraped off using a chisel-ended scraper.
Operation 9.	The hammer was quenched in cold water.

Discussion

The iron was forged into a hammer in a time of 1 hour 46 minutes. The only finishing work necessary was the removal of superficial hammer marks on the two striking faces, which was done with a file. Filing removed 165 g of metal. The finished weight of the hammer was 1.27kg.

It is possible that the design of a Roman hammer may have been influenced to some extent by the size and shape of the billets from which it was formed. The billets from Newstead (Curle 1911) are of a very convenient shape and size from which to forge hammers of this size. Provided the iron was well-consolidated, a very good hammer could have been made with a minimum of forging.

A hammer made for forging hot iron does not need a hardened striking face, as the metal being struck is considerably softer than the hammer. If the metal being struck is cold, and therefore as hard or harder than the hammer head, hardening is necessary.

The hammer was subsequently used for a series of experiments to reproduce Roman iron artifacts. It was given no special consideration and was subjected to the same treatment as any other hammer used in the forge.

The Roman hammer used as a model for this experiment differed from modern hammers in that its eye (the hole into which handle is inserted) is shaped like the frustum of a cone, as shown in the top drawing in Fig. 4.3. In use it soon became obvious why this type of Roman hammer was hafted in the way shown in the bottom drawing in Fig. 4.3. During hammering the head of the hammer moves forward and this forward motion forces the head of the hammer against the wedge, thus locking it. With a tapering hole in the head, the handle of the hammer has to 'protrude beyond the head for the same function to take place. In modern hammers, the eye is a bell-mouthed hole at both ends (Fig. 4.4).

In use, this hammer had a very different feel from a contemporary hammer, but it is well balanced, and a steady hammering rhythm was easy to maintain.

Using a Vickers hardness testing machine hardness tests were conducted in three locations, one millimetre from the striking face, round the eye hole and at the peening end of the hammer head and values of 89VPN, 90VPN and 89VPN respectively were recorded.

No attempt was made to harden the head and, as the hammer was used in the experiments a mushrooming of the head took place. This was remedied by cold forging the striking face back to its original shape which was easy to do and carried out as soon as any spreading had taken place. After three such rectifying procedures a further hardness test was carried out one millimetre from the striking end, this gave a VPN of 142 showing work hardening had taken place.

When used to forge iron at normal forging temperatures, the reproduction Roman hammer head showed no spreading on the striking surfaces, but when it was used on cold iron, the head showed considerable mushrooming after a short spell of working. The same mushrooming effect can be seen on hammer heads from Germany; Pietsch (1984) and Gaitzsch (1980), and in England, Hod Hill (Brailsford 1962), and indicates that these hammers may have been used on cold metal. Modern hammers are made from alloy steels. They are heat treated in such a way that the body of the hammer is very tough and the striking surfaces are hard.

In antiquity, the major ferrous metal was wrought iron (Coghlan 1977) and it is probable that most hammers would have been made of this material. In its normal condition, wrought iron does not contain enough carbon to allow hardening. Carburising is the only way to make a hard striking face on wrought iron. A case-hardened coat of 1 mm will take in the region of six hours heating time. Only the striking surface will have been hardened, and the main body will still be soft and subject to the same distorting forces as unhardened wrought iron.

There are advantages in not hardening a hammer head. When any deformation takes place, the head can easily be corrected with a few blows from another hammer and restored to full working condition. A fully-hardened steel hammer-head may have a longer working life than one which is left soft, but remedial action would require the services of a skilled blacksmith, and it would take him considerable time to restore any damage.

Conclusion

Hammer heads made from wrought iron have definite advantages if left unhardened. They can be produced from standard billets with a minimum of forging, and a minimum of filing to clean them up. When forging hot metal, an unhardened hammer head is just as effective as one that has been hardened. Any mushrooming can be repaired very quickly.

The shape of the eye may have been to do with speeding up production. A bell mouth hole is two frusta back to back and requires more time to produce than a single tapering hole.

It has always been assumed that hardening of iron artifacts was not extensively carried out in the Roman period because of their blacksmiths lack of understanding of heat treatment. It is possible in some cases that hammers may have been deliberately left unhardened.

It is possible that the Roman blacksmiths had found that hardening of some tools and weapons offered no advantages, rather than that they lacked the technology to harden them.

4.4 The production of Roman styli by forging and machining on a pole lathe

"Styli were made of various materials: bronze, iron and bone, and those of iron were probably among the cheapest. The fact that they were forged not cast, as were the bronze ones, means that the addition of decoration was time consuming and, in consequence, is relatively rare, although corrosion has probably destroyed what decoration there was in some cases" (Manning 1985a, 85)

Iron styli are relatively common and several types have been identified and classified (Manning 1985a, 85). They were produced over a considerable period of time and were an essential part of everyday life to many Romans. Although most iron styli are in the corroded condition described by Manning, those found in the special anaerobic conditions at Vindolanda are well preserved and show a considerable amount of applied surface decoration. A study of these styli has shown that the decoration may have been applied by the use of a lathe rather than by the use hand tools. In order to establish a possible method of manufacture and assess the time needed to produce iron styli, a set of experiments was conducted to make a copy of a Roman stylus from Colliton Park, Dorchester, Dorset (collection Ref: 1697 C; 6.11.37; lab number 127; see Figure 4.5 a.

The original stylus is badly corroded and it is impossible to visualise it in its original condition. Before making a reproduction it was necessary to produce a detailed working drawing, defining both form and dimensions. To make the task easier the stylus was considered in three parts.

The eraser end Figure 4.5 b

The body Figure 4.5 c

The scriber end Figure 4.5 d

Figure 4.5 e shows the stylus complete.

The stylus was carefully measured with a vernier calliper, and drawings were made of each element. These measurements indicated that the existing body of the stylus has a roughly rectangular section, but this may be due to corrosion. Reference to Manning (1985a, Fig. 24) does not show styli with rectangular sections, so it was decided to make the body in the form of an elongated truncated cone (Fig. 4.5 c). As can be seen from Figure 4.5 a, the original stylus has a marked curve. The reason for this curve is uncertain. Robin Birley (pers. comm. 1994) states that similar curves seen in styli found at Vindolanda are the result of damage during excavation, and that most of the styli found at Vindolanda are perfectly straight. Therefore since a curve is not found in all styli, it was decided that the curve was introduced at some time after manufacturer, and did not form part of the original design. To ascertain the original length of the stylus before the curving took place, a length of soft copper wire was bent around the form of the original, marked, then straightened and measured. This information was used to produce a working drawing.

It was not possible to carry out any metallurgical analysis of the stylus to determine the composition of the material because of the damage that would be inflicted. The stylus is magnetic and the surface has a deposit of what appears to be iron oxide (rust) This confirms that the stylus is a ferrous metal, and since it was designed to be used on soft wax, the use of steel that can be hardened seems unnecessary.

Because this stylus is made from a ferrous metal from the Roman period, the only alternatives are wrought iron or steel. Even though the stylus was only used on soft wax, continual use would have eventually dulled the point and it would have needed to be re-sharpened. If the point was left soft, the owner could have easily done this himself, using almost any stone, but to sharpen a hardened point would have required special skills.

Production method

When making an artifact for the first time, the correct sequence of operations is not always apparent. The stylus is made from a single piece of metal, and, it is often the case that several attempts are made before the maker can arrive at a sensible and cost-effective order of procedure. It was decided that mild steel could be used for the first attempts to make a stylus, because at modern prices it is cheaper than wrought iron, and any errors made during forging would not waste expensive material.

A number of objectives were established for this experiment which were to determine:

(i) the minimum man hours required;

(ii) the minimum fuel consumption needed; and,

(iii) the minimum material consumption.

No surface decoration was to be applied.

After forging, the whole surface was cleaned back down to bare metal using a file, rather than any kind of rotating abrasive wheel, in order to simulate the conditions during the Roman period.

Equipment: Forge. Anvil. Straight peen hammer. Hot set. One pair of tongs, Fuel, (charcoal).

Operation 1	Forge scriber end flat.
Operation 2	Forge body of stylus into a taper.
Operation 3	Mark body length of stylus.
Operation 4	Cut stylus to length.
Operation 5	Forge scriber end to a round point.
Operation 6	Remove scale with a file.

Total production time 62 minutes.

The experiment established that the above sequence of operations would produce a stylus that is a replica of the original. The experiment was repeated using wrought iron measuring 25 mm x 6 mm which was then forged down to 6 mm square. The operation was conducted at normal forging temperature for wrought iron (yellow heat, approx 1100° C). It was observed that there was some surface cracking at various places on the bar. It is normal practice when forging wrought iron, to take a welding heat on the end of the bar before it is drawn down. A round cross-section was then forged to form the point of the stylus. After forging, considerable splitting was visible, which could not be rectified with fire welding. This part was unsuitable for further work and was cut off. A second square point was forged, with the same result, and was again cut off.

The splitting was caused by large slag inclusions acting as stress raisers. The size of the inclusions had to be reduced to make a more homogeneous structure. This was accomplished by taking a series of welding heats along a length of 100 mm. The bar was heated to 1200°C and given very heavy hammer blows to reduce the size of the slag inclusions in the iron. A second stylus was forged, but, when it was almost completed, a crack was observed at the junction of the body and the point. Close inspection showed that the crack extended to 90% of the cross-section. Further work was abandoned.

Cracks are often due to slag inclusions and other impurities in the metal. Stress raisers also occur where there is a change in cross-section, i.e. where the scriber end and the eraser end meet the body. To reduce the effect of the stress raisers, the transition should be made as smoothly as possible.

In its present condition the rest of the bar was unsuitable for further work of this nature. The size of the slag inclusions became of greater significance as the cross-section of the bar decreased and the complexity of the forge work increased. In order for the material to reach the level of refinement needed to allow it to be drawn down to small cross-section, it was necessary to expel a greater quantity of slag. The same bar 25 mm x 6 mm was reduced to 6 mm square cross-section, taking a total of 10 fire-welding heats and striking very heavy blows with a 2.27 kg hammer. This took 53 minutes. It produced a much more compact structure which was then successfully forged into a stylus with no evidence of cracking.

The total forging time to make the stylus was 11 minutes 25 seconds. The procedure was repeated to confirm the forging time and quality of the metal. The second stylus was produced in a time of 10 minutes 31 seconds. The forging was then cleaned and filed to its finished shape in a time of 49 minutes. This gives a total time of 59 minutes 31 seconds.

Although the stylus produced was fully functional and was close to the original in appearance, there were some marked differences. The stylus was not truly cylindrical and the scriber end was not concentric with the body.

To determine whether the accuracy of the originals could have been produced by hand tools alone, a number of second year engineering undergraduates were given lengths of 6 mm diameter mild steel and asked to produce copies of the Colleton Park stylus using hand tools. All examples were of the same order of cylindricity and concentricity as the author's own copy. This re-enforced the notion that hand tools alone could not produce the accuracy shown in the Vindolanda styli.

The author had the opportunity to examine some of the iron styli found at Vindolanda and found them all to be truly cylindrical, the scribers being concentric with the body. The surface decoration of different grooves and beads were so similar to the kind of decoration seen on wooden items turned on a lathe, that it was decided to experiment to see if iron and bronze could also be machined on a lathe.

The case for the Romans having rotary lathes (those that rotate in only one direction) has been argued most convincingly by Muntz (1972) and Blagg (1976) but examples of such lathes have not been found in the archaeological record. However pole lathes that impart a reciprocating motion to the workpiece (rotates one way then rotates in the opposite direction) are known to have existed since at least 300 B.C. (Hodges 1971).

On modern metal cutting lathes a considerable amount of force is directed onto the cutting tool point. The pole lathe used in this set of experiments required a force of 250 N to depress the pole (Fig. 4.6). (This force is dependent on the diameter of the pole, the distance from the ground to the point were the pole bends, the material the pole is made from, as well as such things as climatic conditions and the amount of moisture in the pole). It is therefore impossible to put an exact figure on it.

As the author had no experience of hand turning metal, several hours practice were required to master the technique. After experimenting for two days and making numerous mistakes, a length of mild steel was successfully turned into the shape of a stylus using a pole lathe and plane carbon steel cutting tools.

The next set of experiments was to determine the amount of time to produce an iron stylus by forging and machining on a pole lathe. The body and the scriber end of a stylus were forged to shape but the eraser end was not, because this end was required to hold the metal between centres for turning (Fig. 4.6).

Total forging time 7 minutes 17 seconds. Both ends of the forging were centre punched to provide locations for the lathe centres. When the metal was cool, the surface oxide film was removed with a file. The bar was put between centres on a pole lathe and machined to shape with hand held tools and finished to a polish with a sand stone. Total time for machining 41 minutes. Total production time 48 minutes 17 seconds.

This is a saving in time of 11 minutes 14 seconds over the other method (59 minutes 31- 48 minutes 17 seconds) and it produced a cylindrical stylus with the scriber end concentric with the body.

As a comparison, a length of bronze rod was machined on a pole lathe, the eraser end of the bronze rod was forged cold and filed to shape. There was no hot forging and the total time to manufacture the stylus was 51 minutes.

Discussion

Styli have been found on many sites (such as Vindolanda) and it is assumed that they were produced in quantity rather than as one-off specialist items made for individuals.

The scriber end and the eraser end could have been formed between pairs of dies made to produce that particular shape. (In the same way that coins are struck.) This would have reduced the production times to as little as four to five minutes, perhaps less. If the manufacturer had been faced with repeat orders, it would have been economically sound to have invested capital in the making of form tools for forging. There is no hard evidence to support the suggestion that there were specialist manufacturers of styli, although it is a strong possibility that this was the case.

The wrought iron has to be forged to small cross-sections, and the quality has to be of a sufficiently high standard to prevent splitting. This experiment showed that items which have a small cross-section are best made from iron in which the slag inclusions are reduced to such a size that they will not act as stress raisers and cause fracture during forging. Providing the metal is of sufficiently good quality, forging is straightforward and can be carried out with very simple tools. The filing by hand and cleaning time was almost five times that of the forging time. In the first experiment, all cleaning was carried out using hand tools, but this may give an exaggerated indication of the time required. The machining of the forging on a pole lathe not only reduced the finishing time but also produced a stylus that was of the same standard of finish and cylindricity as the Roman originals found at Vindolanda. It is possible that rotating stone wheels could have been used in the cleaning operations, but the use of such wheels in antiquity is not known.

Conclusion

Iron styli need to be produced from high quality iron. The scriber end does not need to be hardened and is better left soft. The manufacturing time could be greatly decreased by the introduction of form tools in the forging operations. The finishing of the stylus after forging can take up to five times as long as the forging. It is necessary to use a pole lathe in order to archive a level of accuracy and finish to the same order as the Roman stylus. Because iron is so subject to corrosion, it is not possible to estimate the numbers of styli that were produced. It is not unreasonable to suggest that large numbers of these items were manufactured all over the Roman world. Under these conditions, it is probable that manufacturers would have introduced whatever methods were available to increase production, thereby increasing their profits. Manning (1985a, 85) has suggested that styli made of iron are among the cheapest of their type, but, unless production times can be greatly reduced from those recorded in this experiment, their cost is likely to have been significantly greater than previously thought.

The time needed to apply surface decoration using a pole lathe significantly reduces the cost of such decoration and the grooves and beads seen on many styli could easily have been applied with the aid of a lathe. Non ferrous metal can easily be turned on a pole lathe and it is possible that some bronze styli were made by turning a bar and not by casting.

4.5 Making Roman Nails

Nails are one of the most common finds on many archaeological sites from the Roman period onwards. On first inspection, a nail does not appear to be a complicated artifact to produce, no more than a point at one end and a head at the other. This notion may have given rise to the misunderstanding that the making of nails is a simple skill.

For most applications, nails are required in very large quantities. In the construction of wooden-framed buildings, ship building, and hob nails for the soles of army boots, huge numbers of nails were used. When large quantities of any artifact are needed, the whole production process has to be viewed as mass production. The term mass production in the modern sense means the production of large quantities of components which are made to a set of tolerances such that they are interchangeable. Here the term mass production means simply the production of very large numbers of items.

The hoard of nails found at Inchtuthil (Angus, Brown and Cleere, 1962) contained a very large number of nails of a nail type described by them as nail type D (see Fig. 4.7 and Table 4.4 for details). Since this nail type was used in such large quantities, it was decided to make a copy of one for the first experiment.

Nail production

The objective of the first experiment was to determine the operation sequence and time required to manufacture a copy of a nail type D from Inchtuthil. Olhaver (1939, Abb.36) gives a sequence of operations for nail making. From the writer's experience of forge work, the sequence is an over-simplification so it was regarded as a guide, rather than a indisputable sequence of operations.

Metallographic examination was carried out on the nails from Inchtuthill and the head of a nail from Group A(ii) was formed by upsetting the shank (Angus, Brown and Cleere, 1962, Fig. 6).

To confirm the method of manufacture was the same for Inchtuthil nail type D, one of the nails was sectioned (see Fig. 4.8). This showed the nail was made from a single piece of iron and the direction of the grain flow between the head and shank indicated that the head had been forged by upsetting the shank. Nails were normally produced from iron made by the bloomery process. The nearest modern equivalent to this material is mild steel. As their working characteristics are similar at normal forging temperature (although different at welding heat), it was decided for reasons of cost (modern wrought iron is four times the price of modern mild steel) to use mild steel for this series of experiments.

The tools required to make nails are quite simple. They are: a forge hearth, an iron bar, a hammer, a hot set (for cutting hot metal) and a nail-heading tool. The hot set is shown in Fig. 4.9 e. (Figures 4.9 a to 4.9 d are dealt with later). It is a chisel, set firmly in an upright position. The hot metal is placed on top of it and then struck with a hammer, thus cutting the bar. Figure 4.9 g shows the nail heading tool, and Figure 4.9 h shows a section through this tool, illustrating a tapered hole. A nail from the Silchester collection fits into the Inchtuthil type D classification. This nail was studied, and it was decided that the following sequence of operations was the most likely method of production. The drawings are not to scale.

Production Procedure

(a) Manufacture a bar of suitable length and cross-section (Fig. 4.9 a).

(b) Heat to bright red and hammer the end down to a short square tapering point (Fig. 4.9 b) [technical term, drawing down].

(c) Heat to bright red and hammer a long square point using the edge of anvil and a round faced hammer (Fig. 4.9 c)

(d) Flatten sides on anvil face with flat faced hammer (Fig. 4.9 d).

(e) Cut off bar with hot set (Fig. 4.9 e).

(f) Figure 4.9 f shows the short square point has been partly formed by the cutting operation.

(g) Nail ready to be positioned in nail heading tool(Fig. 4.9 g).

Figure 4.9 h illustrates a cross-section through the nail-heading tool, showing that the hole is tapering.

(h) Heat to 1000°C. Position blank in heading tool and form head (Fig. 4.9 i).

Figure 4.9 j The finished nail still in the heading tool.

Figure 4.9 k The finished nail.

Nail Experiment 1

A nail from the Silchester collection (ref. 80/30 1989 1809 137) was taken as a model for the first nail experiment. It corresponded to Inchtuthil nail type D (30 mm long, shank size 8 mm, diameter of head 15 mm). The above sequence of operations (Figs. 4.9 a to 4.9 k) was carried out, and a time of 3 minutes 55 seconds to make a nail was recorded. The sequence was repeated twice more, and times of 3 minutes 10 seconds and 3 minutes 8 seconds were recorded. A total of ten nails were reproduced and the average time was 3 minutes 7 seconds per nail. In these experiments the blacksmith worked entirely alone.

Nail Experiment 2

Inchtuthil nail type D was used as a model again. The blacksmith worked with an assistant, whose task was to manage the fire. This involved keeping it at a suitable temperature, ensuring the fire was clean and well stoked with charcoal. The other main duty was to give the blacksmith lengths of iron bar that were at the correct temperature for forging.

The head of the nail has to be forged at a higher temperature than the shank, because the metal has to be at its most plastic to allow easy deformation when the head is being formed by upsetting. This was a demanding task and constant vigilance was needed to see that the blank ready for nail heading was not over heated. Two bars were worked together, one in the fire heating, the other being forged. The sequence of operations was the same as in Experiment 1. It was possible to keep up a steady rate of production, because the fatigue factor was not significantly high, which it had been in the first experiment. Twelve nails were reproduced, and with this system, it was found to be possible to produce a nail in 2 minutes 55 seconds. Two minutes were spent on making the shank and cutting the nail blank off the bar, and 55 seconds for forging the head.

Nail Experiment 3

The blacksmith worked with three bars, with the aid of an assistant. The assistant's tasks were the same as those in Experiment 2. The production times were the same as in Experiment 2, but the extra bar meant that the smith had no rest between each nail, and was hammering continually. After a period of nineteen minutes the smith had to stop because of fatigue and was unable to start work again for twelve minutes. Six nails were reproduced. Production times were found to be the same as those in Experiment 2.

Producing large nails

A nail of Inchtuthil type A(ii) was reproduced. This is a large type of nail whose length ranges up to 27 cm. There are several possible ways to manufacture nails of this size, but two possibilities present themselves as the most likely:

(a) The nails were made from bar that is the size of the head and the shanks were made by forging the bar down to size; or,

(b) The bar was the same size as the shank and the heads were made after the shank was formed.

In experiments 4A and 4B the blacksmith worked alone.

Nail Experiment 4A

A bar 33 mm diameter was forged down to 12 mm square, and tapered to a point, over a length of 241 mm. This took 58 minutes, and was a very tiring procedure. It was very difficult to keep the centre of the head and the centre of the shank on the same axis. This resulted in a head that was off-centre and it was felt that this was not the most efficient method of production.

Nail Experiment 4B

Nail type Inchtuthil A(i) shown in Figure 4.7 was reproduced in this experiment. A bar of 12 mm square was forged down to a tapering point over a length of 240 mm. The bar was cut to length and the head formed in a nail heading-tool. This took 38 minutes. The experiment was repeated, and a time of 34 minutes was recorded. Four nails of this type were reproduced in the same manner as experiment 4B and an average time of 34 minutes per nail was recorded. It seems that experimental method 4B would have been the preferred method.

Production times for nails

The following production times for nails found in the Roman Legionary Fortress at Inchtuthil have been calculated using results from the experiments conducted to make copies of Roman nails. It was considered unnecessary to make copies of all the nail types because the lengths of the shanks are such that the amount of work involved, and thus the time taken, would be almost identical. Table 4.4 shows the size range of Inchtuthil type B overlaps with type A(ii) and the times for B, C, and F, can be estimated from the times to make A(i) and A(ii). Type E can be estimated from the time to make D.

Nail type	Length (mm)	Number of nails in hoard	Experimental time to make each nail (mins)	Time (hours)
Ai	247	A(i)+A(ii) 1344	30.0	672.0
Aii	372			
B	171 to 242	B+C	25.0 *	10453.3
C	102 to 156	25 088		
D	73 to 102	85 128	04.0	5715.2
E	38 to 70	763 840	03.3 *	42011.2
F	216 to 222	28	30.0 *	14.0

*Estimated time.

Table 4.4 Estimated production times for nails.

The time taken to have made all the nails amounts to 58 865.7 hours. Assuming a working day of ten hours (this does not include rest times) then the time in working days is 5 887 days. If the working year was 300 days, then the total time is **19.62 man years of work**

Discussion

It is unrealistic to suppose that a smith worked alone. He would have had the assistance of at least one striker and, in all probability, an apprentice, who would have worked the bellows and seen to the management of the fire. In experiment 1 where the smith worked unassisted the recorded times would necessarily have been longer.

Experiment 2 presents a more realistic production time. With small nails such as types B & C the assistance of a striker would not significantly have decreased the production time but would have reduced the fatigue on the smith and therefore increased the amount of work he could do in any one day. This would mean the production of a greater number of nails in a day, but not have reduced the time taken to make each one. Any fully trained and experienced blacksmith is capable of working for a considerably longer time than an amateur, but there is a limit to the length of time even he can work before fatigue forces him to rest. Under normal conditions, a smith working alone gets a short break between hammering, while he is waiting for the iron to heat up again. Even when working two pieces of iron consecutively, it is possible to take a rest between hammering. Experiment 3 gave no opportunity for a break, and after 19 minutes, exhaustion made a rest imperative. Three bars worked at the same time did not decrease the time required to make a single nail, and caused a decrease in the output per day because of the time required for rest periods. There are other working methods which could be tried, but on balance the time of 3.00 minutes ± 5 seconds seems the optimum for Inchtuthil type B and C nails.

When making large nails such as Inchtuthil type A(i) and A(ii), the services of strikers would have been essential in order to produce quantities of nails. Pitts and Joseph (1985, 301), state that it is "not unreasonable to postulate the existence of communal forging hearths because of the large number of nails".

It was only possible to obtain the services of one striker for these experiments, but simple division will provide a very close figure for production time for any given number of strikers. One objective of these experiments was to discover the most efficient method of production. This does not necessarily mean that the Romans used this method, only that it is the quickest way.

Training the work force

The making of nails is not simple forging but it is straight forward and it is possible that legionaries might have been used to assist in forging nails. To test this hypothesis a set of experiments was conducted to discover how long it would take to teach unskilled labour to make nails. Two girls in their twenties from The Department of Archaeology, The University of Reading and two children, a girl aged ten and her brother aged seven, volunteered to take part in the experiments. The experiments showed that, irrespective of age, the volunteers were able to learn the skills of making the shank of small nails in under a day, however, making nail heads is more difficult and it took two to three days of practice for the volunteers to accomplish it. The experiment showed that unskilled labour, including children, can be taught nail making in a short space of time and that one skilled blacksmith could easily oversee a large number of unskilled workers making nails. All the volunteers had difficulty making nail heads. The heads tended to be offset from the shank and were often distorted. The author has been loaned a small number of nails from Inchtuthil and there are nails in the sample with heads misshapen in the same manner as those made by volunteers. This could be an indication that some unskilled labour was used to make the Inchtuthill hoard.

It would seem to be possible that the making of nail shanks was done by unskilled labour and the heading by more skilled workers. The larger nails Inchtuthill type A(i) and A(ii) require more skill and so would probably have been done by a skilled blacksmith.

Teaching nail making

The second part of this set of experiments was to examine how much instruction was required to enable unskilled workers to produce the simpler parts of nails.

The volunteers stated that they had never been in a blacksmiths forge before and had never done any forge work. Each volunteer was then given a full briefing about the objective of the experiment, followed by a demonstration of each stage in the making of a nail.

The first volunteer was G a girl aged 24 years.

Forging down a short square point was demonstrated. G's first attempt took three heats and developed a fairly long point.

The demonstration was repeated by the instructor.

G's second attempt she forged a short point in two heats. In her third attempt, she again took two heats to forge a point but the quality was considerably improved. Her fourth attempt, again took two heats to forge, and the quality improved even more. In her fifth attempt, she produced a good quality point in one heat.

The total time taken (including demonstrations) was 1 hour 3 minutes. The standard of the points G produced were very good and certainly more than adequate for nail making.

Roughing the shank to shape was then demonstrated. By the fifth attempt G was able to produce a satisfactory shape in two heats. Refining the point demonstrated and G's second attempt was good enough for a nail. G found heading very difficult, and a lopsided head was the best that was achieved. This is not at all unexpected as the skills required to produce a nail head are quite considerable and one would not expect a newcomer to forge work to master the skills without several weeks of practice.

A second volunteer M in her twenties assisted in a similar experiment after a days practice the results were similar to those of the first volunteer. It took her four minutes to make the shank (point and cutting off). The heading took her an average of 15 minutes. She also found difficulty forging heads.

R, a girl aged 10 and her 7 year old brother G, volunteered to try nail making. (They were given the same amount of instruction as the other two volunteers). R was able to produce nail shanks after only two attempts but found putting the heads on very difficult. Her brother G experienced the same level of difficulty.

Conclusion

Some parts of Roman nails (the shanks) were probably made by semi-skilled labour and the more difficult parts (the heads) were probably done by skilled craftsmen. Having found that the making of small nails can easily be done by women and children it seems possible that they could have been used. Women and children could supply a cheap workforce for making small items, freeing men to undertake heavier work.

Although using mass unskilled labour probably does not significantly reduce the production time for nail making, it does mean that very few skilled smiths would have been needed on any one military site and that much of the work could have been done by other personnel trained on the site.

When a timber fort was constructed, the most sensible and practical way for nails to be made was for the iron to be transported in the form of billets, which were then forged into various artifact including nails. Indeed, the supply of iron in the form of billets is attested by the recent discovery of several such billets at Strageath, their weights ranged from 5.7kg to 7.4kg (Pitts and Joseph 1985, 113). Unskilled labour could have been used for the forging down of the billets into bars, and for making the shanks of the nails so that only the heads would been forged by skilled men. It is the usual practice when smithing to keep the metal hot for the whole of the forging, thus conserving heat and reducing the fuel consumption, but this is not always possible.

4.6 The manufacture of disposable weapons for the Roman army

Evidence for the type of weaponry used by both Republican and Imperial Roman armies has, until recently, been scattered through classical art and literary sources, as well as the archaeological record. Many archaeological reports have carried drawings and comments on weaponry from individual sites, and some museum catalogues have identified weapons from their own collections. Bishop and Coulston (1993) have done much to bring this scattered information into a logical sequence of development. There have been many reproductions of Roman military equipment, and fine examples can be seen in the Museum of Antiquities, Newcastle-upon-Tyne and The Roman Legionary Museum, Caerleon, Wales. Most of them are of excellent craftsmanship, but many have been made using modern tools and materials, with no consideration given to production times or methods.

In these experiments the conditions of a Roman blacksmith's forge were re-created as closely as possible. Copies of tools and the accepted technology of the period were used to produce the weapons in this set of experiments. Projectile weapons formed a significant part of the arsenal of the Roman army and those re-constructed in this set of experiments were used and, for the most part, probably manufactured within Britain during the Roman occupation. Only iron projectile heads, fitted onto the end of a wooden shaft are considered here. These can be divided into two categories, those to be thrown by hand (javelins and spears); and those employing mechanical means of propulsion (arrows and bolts). A fighting force, using such weapons, had to be assured of a constant and reliable supply of replacements: there could be no assurance of recovering those expended. Those disposable weapons which were made and used with the expectation of non-recovery are examined in this part of the book.

Weapons would have been required in large numbers, as Wheeler (1943, 281) has suggested. The archaeological record contains a large number of excavated examples. This section will concentrate on those ferrous weapons known to have been used by the early Imperial army. Using a set of experiments, three questions will be addressed.

(1) What were the likely methods employed by Roman armours to produce disposable ferrous weapons?

(2) How long did such weapons take to make?

(3) How much material, including iron and fuel, was consumed in the maufacture?

Modes of production

There are three principal systems of producing iron weapons from billet or scrap:

(1) Production by the legionaries and auxiliaries, either in camp, or in the field (Bishop 1985);

(2) The imperial *fabricae* of the late Roman empire (James 1988);

(3) Production by small individual workshops of special items (MacMullen 1960).

Method of manufacture

There are three fundamental methods of manufacturing items from metal: casting, machining away the unwanted material, and forging. The casting of iron was not practised by the Romans (Tylecote 1987), and machinery for the removal of large quantities of metal was not available at this time. By a process of elimination, the only technique left is forging. With any forging operation there are always several paths which can be followed. They may all result in the same end-product, and artifacts made by these several methods may be indistinguishable from each other. A set of assumptions was therefore developed to govern all the items made in this set of experiments:-

(1) All the items were made as if they were mass-produced.

(2) The blacksmith worked with only one striker. In practice, a blacksmith can work with up to three strikers. The blacksmith used a 2 kg hammer, and the striker a 10 kg sledge hammer.

(3) Wrought iron was used, as there is no evidence for the use of steel for disposable weapons.

(4) The iron was received in billet form to be forged into appropriate bars or sheet, in order to make blanks from which the weapons could be forged. It is also assumed that there was minimum of material wastage.

(5) The items had to be produced in the minimum amount of time, with the minimum of fuel consumption.

(6) Production had to be such that individual operations could be performed by semi- skilled labour.

(7) No items were hardened.

(8) Charcoal was used as a fuel but there is strong evidence of coal being used for forging.

As no evidence has so far come to light for the use of mechanical hammers in the Roman period, it has to be assumed that manpower supplied the energy to refine the billet. It was common practice until recently for a blacksmith to have at least one striker working for him and there is pictorial evidence to show the use of strikers in antiquity (Healy 1978, 164 fig 52). Indeed, it is not possible for one man alone to perform many of the different operations used in blacksmithing such as swaging and fullering. The number of strikers is open to question.

Figure 4.10 shows three strikers at work. The shape and size of the anvil will determine the number of strikers who can physically fit round it and it is possible that as many as three strikers could be employed in Roman times to refine blooms. Striking is a skilled task, but it is a skill required in a very specific and narrow area, and can be taught quickly. It does not compare with that of a blacksmith, who has to have a wide area of expertise, and considerable time to learn. A number of strikers can be trained as a team in a very short space of time; two to three days would be quite sufficient. At the end of this period they will be able to work efficiently under the orders and direction of a blacksmith. Therefore it follows that forging billets into specific cross-sections is a skill which can be easily taught.

It is possible that when an army was in the field, there were *immunes* who were blacksmiths, but that others dealt with forging bars to the size necessary for the making of specific weapons. The latter task would have been the responsibility of the more skilled smiths. During the experiments only one striker was employed.

It will be noticed that in the early stages the amount of heating time is substantial, because of the large volume of metal being heated. During heating time the workforce is idle, so it is possible that the team would work with two bars; one being heated, while the other was being forged. In this way a more efficient use is made of manpower. It is unlikely that any team would work on more than two billets at once. Striking is very hard and tiring work and a striker, no matter how strong, will soon become exhausted by continual work without rest periods. If the striking team is given a short rest between hammering sequences, it will be able to work for a much longer period without fatigue. Although it is possible to work the whole billet at once, it was probably not the common practice to do so. The billet as a whole is heavy and very difficult to manipulate. Usually it requires a considerable time to bring up to forging heat and, although this time is reduced during subsequent heats, the heating still consumes a large quantity of fuel and time. Although there is no direct evidence of billets being cut up, evidence from the 15th century (Lapsley 1934, 514) shows that it was common to supply the smith with billets, about one third the size of the Roman billet. A combination of ergonomics and common sense show that a billet of 2.5 kg is much easier to heat and handle than one three times the size.

Summary

Iron was probably supplied to the Roman army in the field in the form of billets. The quality was sufficiently high for the smith to manufacture artifacts straight from the billet. The latter was smithed down into strips which provided blanks for the subsequent working. The forging of billets into bars is a semi-skilled operation, and operators could be trained in a short space of time. The smithing of the billet was carried out by a smith and up to three strikers, and not more than two billets were worked by any one team. The preparation of blanks consumes a large quantity of both fuel and manpower.

4.6.1 The manufacture of a pilum

There are various types of *pilum* from the Roman period (Schulten 1911; Schulten 1914). However, the model for this experiment was supplied by reference to Manning (1985 a 75, Fig. V22). In all the following experiments the objectives were to determine:

(a) the amount of time required,

(b) the metal and fuel consumed

(c) the most likely production sequence.

Given the technology available, there are only two possible ways of producing a pilum. The whole shape was made by cutting it from a solid bar, using saws and files, or it was forged from a bar and finished with files and abrasive stones. The first method would consume an enormous amount of time and be very wasteful of material. Hand methods would not be able to meet the needs. Therefore, the most likely method of producing a pilum is by forging followed by filing. The finishing would probably have been done with hand-held files, and abrasive stones. The existence of a rotary wheel for metal removal cannot be proved at this time.

The first operation was to reduce a block of wrought iron to a bar 685 mm long by 13 mm square. This is enough material to make one pilum and is long enough to handle when forging, without needing to use tongs. The end of the bar was heated to a light yellow heat (1000°C) over a length of 90 mm. On removal, a length of up to 40 mm at the end was quenched in cold water and the end was upset (Fig. 4.11 a). A total of 8 heats were needed to complete the upsetting. The section that was upset was drawn down to a square point (Fig. 4.11 b). Using a fuller and cross-peen hammer the section behind the head was reduced in section (Fig. 4.11 c). The fullering was evened out with a flatter (Fig. 4.11 d). A large radius was left under the head so as not to weaken it during the fullering. The tang was produced by fullering and the use of a flatter. Using a pointed punch, and then a tapering drift, the two holes were made in the tang. The radius under the head was removed by filing to provide a square shoulder. The forging was brought to size by filing, and the finished surface was produced with various grades of file (Fig. 4.11 e). It was decided to stop when the finished item had a clean finish and presented a pleasant appearance to the eye. If a perfect specimen was required, at least an additional one hour for filing would be necessary. Table 4.5 gives weight loss and fuel consumption for the manufacture of a pilum.

Starting weight	874 g
Weight after forging	701 g
% weight loss due to forging	19.79%
Weight after filing	574g
Total % weight loss	34.3%
Fuel consumption	3.5 kg

Table 4.5 Weight loss and fuel consumption for the manufacture of a pilum.

Times

To reduce the billet to a bar: 4 hours 38 minutes (one blacksmith and one striker).

To forge the pilum to shape: 2 hours 7 minutes (one blacksmith working alone).

To file the pilum to shape: 3 hours 41 minutes (one man working alone).

Total time 10 hours 26 minutes.

Discussion

The production of this weapon presents no technological problems; it is straightforward forge-work, of a fairly simple kind. It requires a heat source in the form of a hearth, probably with bellows. A hammer, anvil, swages, (see glossary) and a punch are the only tools needed for the forging. All these tools are known from the Roman period and would have been essential equipment in any blacksmith's workshop (Manning 1985a). The weapon was forged to as near the finished size as possible in order to minimise the amount of time-consuming work involved in finishing by filing. Even so, a considerable amount of metal was removed to bring about the finished item. It is assumed that, as this weapon probably had only a short life span, polishing was not an appropriate process. It would fulfil no practical purpose, but would only add to the cost. This does not mean that legionaries did not polish their own pila, in order to enhance their appearance when on parade. The configuration of the head is such that the point is very strong and sharp. Since it will penetrate most materials with ease, it would not be necessary to harden the points. The design is such that when the pilum strikes an object it will penetrate and when the forward force is expended, the bending moment caused by the weight of the wooden haft will cause the long tapering shaft behind the head to bend. Caesar describes pilum heads bending and rendering the shields of the enemy useless, Caesar, trans. Handford, (1982). As with any thrown weapon, it must be expected that during a battle significant numbers will be lost completely. Of those that are recovered, some will be damaged beyond repair. Others can be easily straightened out for reuse, providing the bend has not caused any surface cracking of the metal. The unhardened points would be easy to re-sharpen. It is likely the irreparable pila would be recycled, as the material would be of good quality and the blacksmith would not be likely to have let it go to waste. A production time of 10 hours 26 minutes for each pilum, together with a fuel consumption of 13.5 kg shows that this was not a cheap weapon to produce. As this type of weapon was produced over a considerable period of time and passed through several minor modifications in design, one is led to the conclusion that the effectiveness of the pilum justified its expense (Bishop and Coulston 1993).

4.6.2 The manufacture of a bolt head (square pyramid)

The catapult bolt-head for this experiment is based on an example illustrated by Manning (1985a, V144, pl. 80). This is a common type of bolt-head with a pyramid-shaped point and a tapering socket. A length of bar was heated to a yellow- red heat (950°C) flared into a fantail shape (Fig. 4.12 a). A further heat was taken to the same temperature as before, and the head was cut off the bar with a hot set (Fig. 4.12 b). In one heat, the pyramid point was forged to shape (Fig. 4.12 c.). Using a bottom swage and the cross peen end of the hammer, the socket was partly formed, and then closed over on the anvil using the hammer. Figure 4.12 d shows the completed head. When the metal had cooled down to black-red heat, the point was hammered to impart a smooth surface finish. The point was quenched. This causes the surface of iron oxide to become hard and imparts a slight case-hardening effect on the surface of the metal. Three bolts of this type were made.

Total number of heats taken during forging 8

Average time to forge bolts 17 minutes 31 seconds.

Starting weights of bolts ; 66 g 69 g 75 g

Finished weights of bolts; 60 g 61 g 70 g

It takes 34 minutes 45 seconds to make the material for one bolt head, and a further 17 minutes 31 seconds to forge it to shape. This gives a total time of 52 minutes 15 seconds. The forging process was repeated twice more when times of 17 minutes 01 seconds and 17 minutes 09 seconds were recorded.

Discussion

The method of manufacture described above is very economical in the consumption of metal. There is no waste from cutting, and the only metal lost is due to oxidation from forging. Cleaning of the surface is unnecessary and a smooth surface can be obtained by hammering when the iron has cooled to black-red heat (750°C) followed by quenching. The resulting oxide film will give some protection from rust. These factors seem to indicate that economy of material may have been a factor in the design of these bolt heads, which were, after all, only expected to be used once. The design and method of manufacture of these bolts suggest that the manufacturers found the quickest and most economical method of producing these items.

4.6.3 The manufacture of a flat-bladed bolt-head

The flat-bladed bolt-head for this experiment is based on an example illustrated by Manning (1985a, V268, pl. 85) and Wheeler (1943, Fig. 93, 3-13). These are triangular or leaf-shaped blades with flanged sockets. The one chosen as a pattern for this experiment has a length of 7.8 cm and a blade with blunt edges. The flanged socket is carefully made. The fact that the edges are not sharpened is explained by Manning (1985a, 176) "such thin blades could be expected to penetrate even if the edges were not sharpened". A close study and a few preliminary sketches of the drawings showed that the bolt had been made from a single triangular piece of metal. The details were arrived at from the text and by scaling up the drawing.

Sequence of operations

Using a hot set, a triangular blank was cut from a strip 7.8 cm wide (Fig. 4.13 a). The cut edges were chamfered from the cutting, but this was removed with a coarse rasp. Using a bottom swage and a cross peen hammer, the flange was partly formed (Fig. 4.13 b and Fig. 4.13 c). The socket was close up around a tapering mandrel (Fig. 4.13 d).

Starting weight of blank:	38 g
Finished weight of head:	33 g
Cutting blank:	2 minutes 7 seconds.
Forging head:	6 minutes 31 seconds
Total production time:	8 minutes 38 seconds.

A second bolt head was made and the forging time was reduced to 5 minutes 12 seconds. With practice at this particular item, a production time of under 5 minutes per item is possible.

Discussion

Bolts fired from Roman catapults were fired with considerable force and, if they struck a human body, would make a deep penetration wound. The shape of the wooden shaft would probably prevent the whole projectile from passing through the body, but would leave the head buried very deep. If the shaft was still intact the bolt could be recovered from a body. If the bolt missed the target, then it would bury itself in the ground, as modern target arrows do. If it struck a stone, the point would probably be damaged beyond repair. This suggests that bolts fired from Roman artillery were probably expected to be used only once. The flat bladed bolt would seem to have to have two major advantages to its design, effectiveness and cost. From the drawing (Fig. 4.13 a) it is possible to see that if a strip of wrought iron 7.8 cm wide was made, then triangular blanks could be cut out in such a way that the amount of waste material would be negligible.

The production time for these items is low, a rate of 12 per hour per man would be possible. They require no finishing by filing and, if treated with wax or tallow, could be stored with little or no deterioration. The blacksmithing skills required could be taught quite quickly and a high degree of proficiency would be reached in a short space of time. Wheeler (1943, 281) asserts that this type of bolt head was "mass produced on Roman sites, probably for use with ballistae". These items could have been made in the field by legionaries.

4.6.4 The manufacture of a spear head

The head for this experiment is based on an example illustrated by Manning (1985 V123 pl 80. Figure 4). With any manufacturing process, there are a large number of possibilities concerning the steps in manufacture. In this case, there seem to be two distinct methods that recommend themselves, on the grounds of minimum consumption of material.

Method 1: A sheet 250 mm x 5 mm thick of an undetermined length was cut up into triangles as shown in Fig. 4.13 a. This had the advantage that, once the sheet was produced, there was no wastage from cutting, and the production of a single item could be very quick. The disadvantages are the time taken to produce the sheet and the very high risk of cracking, due to the slag inclusions present in wrought iron. When wrought iron is being formed into sheets, it can only be worked in one direction. This causes the slag to spread into flakes. These have a small cross-section but a large surface area. The slag inclusions prevent the iron from welding and act as stress risers. As the cross-section of the iron decreases, cracks will form, often near the centre of the sheet.

Method 2: A strip 50 mm x 5 mm x 250 mm was cut and flared to shape. The cross-section of the blank is larger, and can be worked in two directions, so the risk of splitting is reduced. The disadvantage is that material is lost in cutting. (This does not have to be a drawback, because the off-cuts can be re-welded into another billet.) It was decided to use the second alternative for this experiment.

Sequence of operations

The bar was heated to a bright red (900°C) and cut to a triangular shape. When working on thin section material such as this, it is better to keep the forging temperature low, because it is easy to burn thinner metal. After cutting, most operations were carried out at dull red heat. The cutting edges were forged to shape but not to a sharp edge. The material for the socket was flared using the cross peen of the hammer, and the irregularities were smoothed out with the flat face of the hammer. At this stage a hole was punched through the flared end. Using a bottom swage and a cross peen hammer, the socket was started (Fig. 4.13 b) and finally brought to shape over the anvil. As the finishing stages of forging had been performed at a low temperature of approximately 680°C, the surface was quite smooth and had light pits in it. This reduced the amount of time needed to file the surface clean and sharpen the edges. Table 4.6 gives weight loss and fuel consumption for the manufacture of a pilum.

Starting weight:	486 g
Weight after forging:	358 g
Weight after filing:	316 g
Forging time:	45 minutes 28 seconds
Filing time:	1 hour 19 minutes

Table 4.6 Weights and times for making a spear

Discussion

Gilmour and Tylecote (1986, 242) describe the spear heads which they examined as unhardened. The spearheads in the experiments were left soft. If spears are in a hard condition, this could have disadvantages for the legionary in the field. When in the unhardened state, such a weapon can easily be repaired in the field by its owner in the same way that mowers' anvils were used to repair damaged scythes. Weapons practice would also damage projectile points and a unhardened head is easy to repair because any blunting of the edge or point can easily be put right with a sharpening stone. If the edge is hardened, a skilled smith would have to be employed to repair any damage. A spear of this design offers great advantages to the manufacturer and the soldier alike. The manufacturer can produce large numbers of identical items in a short time, there is little waste material and no complicated tooling is required. Much of the work of preparation of the blanks can be done by semi-skilled labour and time-consuming processes such as hardening are eliminated. The soldier can carry out simple maintenance of his own weapon.

4.6.5 The manufacture of a fire-arrow

The fire-arrow head for this experiment is based on an example illustrated by Robertson et al.(1975, 32 14) and Manning (1985a, 177). This is an arrow head in the form of a cage (Fig.4.14). It is made from three lengths of iron, welded at each end. One end forms the point, while the other is fixed into the shaft. The inside of the cage was filled with incendiary material. Almost certainly these arrow heads were used only once, and economy would no doubt play an important role in their design. They would need to be of a sufficiently large size to hold a considerable quantity of incendiary material. Some of the latter would burn up during flight, due to the forced oxygen introduced by the passage of the arrow through the air. There would, however, have to be enough left at the end of a flight to start a fire.

Sequence of operations

Three different methods of manufacture were attempted to determine the most likely production process, and time.

(1) Three pieces of iron, 6.0 mm in diameter and 90 mm long were fire-welded at each end and the welds were drawn down into points. It was impossible to open the bars to form the cage and when viewed the whole structure seemed too small. No further work was carried out on this head.

(2) Three pieces of iron, 6.0 mm in diameter and 120 mm long, were welded at one end. They were opened out and a short length of bar was placed in the centre of the bundle. The other end was welded. The loose piece enabled the cage to be formed, but again the hole was too small.

(3) Three pieces of iron, 6.0 mm in diameter and 140 mm long, were fire-welded at each end and the welds were drawn down into points. The bars were given a clockwise twist of 270°. This twist was then turned anti-clockwise to open it out and form a cage. The amount of material required was 420 mm x 6 mm diameter for each arrow head. Reference to table 4.3 shows it will require 149 minutes to forge enough material for each head from a billet.

Discussion

It seems, on balance, that the third method offers the best method of production. The forging time was 23 minutes, and a further 7 minutes were taken to file the point and the shoulder end to fit in the arrow shaft, giving a production time of 30 minutes for each arrow. A production time of 30 minutes for each item seems very high, taking into account the time to forge the bar from a billet it gives an additional production time of 179 minutes (almost 3 hours). Presumably this production time can be justified by the efficiency of the weapon.

4.6.6 The manufacture of an arrowhead (spade shape)

A close study and a few preliminary sketches of the drawings showed that the arrow head had been made from a single triangular piece of metal (Wheeler 1943, Fig. 93, No. 13). The details were obtained from the text and by scaling up the drawing.

Sequence of operations

The method of manufacture is the same as for the flat bolt heads,except the blank is smaller (Fig. 4.13).

(1) Using a hot set, a triangular blank was cut from a strip 5.3 cm wide. The cut edges were chamfered from the cutting and this provided a preliminary edge. The total time taken was 2 minutes 18 seconds.

(2) Using a bottom swage and a cross peen hammer the flange was partly formed (Figures 4.13 b and 4.13.c).

(3) The socket was then closed up around a tapering mandrel (Fig. 4.13 d). The edges were forged down to as sharp an edge as possible.

Starting weight: 20 g

Finished weight: 17 g

Time: 5 minutes 31 seconds.

Total production time: 8 minutes 38 seconds.

With practice, this particular item could be produced in under 5 minutes.

Discussion

If a strip of wrought iron 5.3 cm wide were made, triangular blanks could then be cut out in such a way that the amount of waste material would be negligible. As with the flat bolt heads, the production times for these items is very fast - 12 per hour per man would be possible and economical because it uses the minimum of metal. They require only honing to sharpen the edges and, if treated with wax or tallow, could be stored with little or no deterioration. The blacksmithing skills required could be taught quite quickly and a high degree of proficiency would be reached in a short space of time. These items could have been made in the field by legionaries.

4.6.7 General discussion

Since all these weapons were expendable, it seems there are two important criteria to take into account in their design: efficient performance and production costs. With the exception of the fire arrow, all the weapons made in this series of experiments were forged from a single piece of material, using the minimum amount of metal. These weapons are made entirely without joints. As these are always a source of weakness, their elimination may have been a deliberate factor in the design. Production is in two stages. In the first the raw material, be it billet or scrap, is forged into a suitable blank for manufacturing into weapons. In the second, the weapons are produced. These items are all simple to make and if the manufacturing was divided into individual tasks they could be made by semi-skilled labour. Certainly blanks for all items can be prepared by semi-skilled labour. It seems likely that these items were made in large quantities, and then stored ready for use. This would be the case during time of war.

The storage of iron items in such a way as to prevent rusting had always been a problem. If iron is left as forged, the oxide coat will give a limited amount of protection. Methods of preventing rust were known in antiquity. Pliny trans. Bostock and Riley (1857) suggests such things as liquid tar, bitumen, red lead, white lead and gypsum. Polished weapons would rust much more quickly. Unhardened iron can be sharpened to a razor edge, or a needle-sharp point. The purpose of hardening is to preserve and extend the effective life of a cutting-edge or point, but this is only necessary when the edge or point is to be used more than once.

Hardening is a process which is carried out to provide a long-lasting cutting edge, and is not necessary on a disposable weapon. Gilmour and Tylecote (1986, 242) state that the examples of Romano-British spearheads which they examined showed no evidence of hardening. The configuration of the head of the pilum and the pyramid-bolt head are such that they will penetrate almost any material. Thus a hardened point would not increase their effectiveness, unless they were to be used against very thick plate armour. Left soft, heads can be repaired in the field by soldiers with little training in metalwork. All the weapons made in this set of experiments were designed to be only used once in combat. During a battle, significant numbers would probably be lost completely, and this probably had a direct influence on the design. It is conceivable that some weapons were recovered intact, which could be re-used with little or no work, while those that were irreparable would probably be re-cycled. When manpower, metal, and fuel consumption are considered, most of these weapons cannot be considered cheap to produce. It must be supposed that the effectiveness justified the expense. It is the opinion of workers who have reconstructed ballistae (Anstee pers. comm. 1993) that a range of 300-400 metres is quite possible. A range of 100-150 metres is possible for a bow and arrow and a range of 30 metres for a pilum. An enemy would be under attack from projectiles for a distance of 400 metres and perhaps further. The pyramid bolt head was possibly designed to be armour-piercing and the flat bladed spear, bolt-head and arrow-head would produce great haemorrhaging wounds.

Summary of manufacturing times

Item	Loss from forging % of starting weight of blank	Loss from filing % of starting weight of blank
Pilum	19.79 %	18.12 %
Spear	26.34 %	11.73 %
Javelin	26.35 %	11.73 %
Arrow flat	13.00 %	1 %
Bolt pyra	20.00 %	None
Fire arrow	27.47 %	1 %
Flat bolt head	None	13.15 %

Table 4.7 Material lost during forging and filing to shape.

Table 4.7 shows that there is considerable loss of material during forging on all items. There is also a large loss of material on the pilum and spear caused by filing. On the other items, weight loss due to finishing is negligible or non-existent. The flat-bladed bolt-head, the arrowhead and the spearhead and the javelin are all manufactured by the same method (see Figures 4.13 a to 4.13 d).

Table 4.8 summaries the production time of the various weapons considered in this section. It can be seen the pilum takes considerably longer to make than any other item in this range of weapons, representing at least one whole day's work. The time involved in finishing both the pilum and the spearhead is substantially longer than the forging time. It is possible that finishing was not carried out by the blacksmith but by other craftsmen who specialised in that kind of work. Both the pyramid-shaped and flat bolt-heads represent the most efficient use of the forging technique, as they are very quick to make and requires no finishing by filing. The only losses of material are due to forging. When the forging times (in the second column of Table 4.8) are considered it shows that all these items consume a large amount of time and fuel just to prepare the blanks before their manufacture into weapons. These times are taken from only two men working on the billet, and only one billet being heated at once. It has been suggested earlier that as many as four men (blacksmith and three strikers)might work two billets together, so reducing the blank-forging time considerably. When the forging times in column 2 of Table 4.8 are considered, we can see they are very fast, in particular the flat bolt-head and the arrow. These items could have been made in the field by legionaries. When making the fire arrow there is a high material loss due to forge welding, but it is essential to the manufacture of this item. There are no refinements, decorations, or anything superfluous in the finish of the weapons. They have only the absolute minimum to allow them to perform their task efficiently. This suggests that their consumable nature had a direct influence on the design. The total production times for some of these weapons indicated a considerable investment in time and materials, and it is suggested that the efficacy of the weapons justified this investment.

Item	Forging time billet to blank h : m	Forging time h : m	Finishing time h : m	Total Time h : m
Pilum	4 : 38	2 : 07	3 : 41	10 : 26
Spear	0 : 54	0 : 46	1 : 19	2 : 59
Bolt hd-fl	0 : 31	0 : 05	None	0 : 36
Fire arrow	2 : 29	0 : 23	0 : 07	2 : 59
Bolt pyra	0 : 35	0 : 18	None	0 : 53
Arrow flat	0 : 30	0 : 05	0 : 02	0 : 37
Javelin	0 : 54	0 : 45	1 : 19	2 : 58

Table 4.8 Forging and finishing times for disposable weapons.

Summary

The recovery of any weapon operating as a projectile is always in doubt and so it is reasonable to suggest that the weapons discussed in this section were made to be used only once. This study has shown how they could be produced using the minimum amount of metal necessary and from this it is possible to infer that the methods of manufacture available at the time affected the design of the weapons. The iron was first formed into blanks from which the weapons were subsequently forged. As this operation is very time consuming, it is possible that some of this work was done by unskilled labour. The time needed to produce large numbers of each weapon leads to the conclusion that the forging was conducted by blacksmiths assisted by a number of strikers.

4.7 Pattern-welded sword

The main reason for this experiment was to determine the exact number of man hours involved in the production of one pattern-welded sword. Fire welding (see glossary) is probably the most difficult skill for the blacksmith to learn, and in modern ironworking it is usually restricted to single welds. Pattern welding involves taking a series of consecutive fire welds on a long length made up of several iron rods or strips. There is no parallel in modern forge-work, and it is not an easy process either to learn or conduct. The earlier experiments were conducted to expand the skills of the researcher. The pattern produced on a pattern-welded sword comes about because irons with differing carbon content are welded together, and when the sword is treated with acid, the different types of iron etch in different ways, thus producing a pattern. Gilmore and Tylecote (1986, 246) have shown that there are at least eight combinations of iron types, and each combination will produce a unique pattern (Fig. 4.15).

The sword in this experiment is made of a core of three bundles, with two strips one on each edge to provide the cutting edges (Fig. 4.16).

If each individual bar is twisted in six different places along its length before they are welded together, the position of the twists relative to each other and the hand of the twists can be varied to give a possible 262 144 combinations.

Tylecote and Gilmore (1986, 246); see Fig. 4.15 show eight different configurations, but in fact the possible number of combinations is enormous, and one craftsman, working on his own, would be able to make only a very small fraction of all possible combinations of swords in his own lifetime. At present there is insufficient evidence to determine if pattern welding was done for decorative purposes or strength (Lang and Ager 1989), but it is a strong possibility that it was mechanically necessary, because of the poor quality of the iron produced by the bloomery process. The author has examined swords from the Nydam collection (Engelhardt 1866) and has reached the conclusion that both points of view are valid. In some instances, the layer of pattern - welded material is so thin that it can only be a decorative feature. In others, the whole thickness of the blade is pattern-welded and this might increase the strength.

The making of a pattern-welded sword can be considered in the following stages:

(a) Raw bloom iron is formed into a billet;

(b) The billet is forged into thin strips;

(c) The strips are forge-welded into a sword blade; and,

(d) The blade is filed into shape

The sword was made to determine the following.

(a) weight loss due to forging

(b) Material lost in finishing

(c) Times for the various operations; and,

(d) Metallurgy of the blade

Materials: Mild steel and carbon steel EN42J

Sequence of operations.

Operation No 1 Both materials were made into strips 505 mm x 10mm x 1 mm.

Operation No 2 Three bundles were made up of alternate strips of mild steel and EN42 J.

Operation No 3 All bundles were wired together.

Operation No 4 Heat 100mm from one end to a bright red and twist one quarter of one turn(90) clockwise. Continue taking heats down the bar until the whole length of the bar has been twisted. Repeat until the bar has been given four complete twists.

Operation No 5 Repeat operation 4 on next bundle.

Operation No 6 Repeat operation 4 but turn anti-clockwise.

Operation No 7 Fire weld down the length of each bundle.

Operation No 8 Wire the bundles so that the anti-clockwise bundle is in the center and fire weld the bundles together.

Operation No 9 Fire weld the cutting edges onto the pattern-welded strips.

Operation No 10 Forge the tang to shape.

Operation No 11 Forge the point to shape.

Operation No 12 File the blade to shape.

Operation No 13 Polish all file marks out with stone hones.

Operation No 14 Case harden.

Operation No 15 Heat blade to bright red and quench.

Operation No 16 Re-polish.

(a) It takes 62 hours 45 minutes to refine the bloom into a billet.

(b) It takes 45 hours to forge the strips.

(c) It takes 7 hours 23 minutes to weld the sword together.

(d) It takes 10 hours 15 mins to file the sword to shape.

One smith working alone will require 125 hours 23 minutes to make a pattern-welded sword of three bundles and two welded-on edges.

Discussion

The amount of time involved in making a pattern-welded sword is enormous. If a working week of 60 hours is hypothesised, then a smith working alone will only produce 25 swords a year. This seems unlikely, especially as a skilled craftsman does not enjoy the same status as an artist and the level of wages he can command is not high. It does not seem likely that an independent smith working alone would be able to produce enough swords to provide a living for himself and his dependants. The level of expertise demanded in producing a pattern-welded sword is high, but should be within the level of skills of any competent blacksmith. The swords examined by the author show that the skills exercised by the craftsman were very high, indicating the sword smith was a specialist at pattern-welding. The possible alternative is that the smith only carried out the welding. Pre-forging and subsequent finishing might be put in the hands of other craftsmen. A possible alternative model for sword production may be as follows.

Raw bloom iron forged into billets

This can be done by semi-skilled labour; the billet is delivered to the smith.

Billets forged into strips

This is not a skilled operation and could be conducted by a team of as many as four men working together on a single anvil. They could all be semi-skilled. Clemetson (1993) has suggested a rolling process for the making of sheet for scale armour. A similar process could have been used for the production of strips for pattern-welded swords.

Strips forged into a sword blank

This involves the forge welding of the strips and is a highly skilled task; this would have to be performed by a skilled blacksmith.

Sword blank filed and polished to shape

This work would be carried out by a cutler who is considered a highly skilled craftsman. The cutlers skills rank alongside those of the blacksmith and he may have been a specialist in his own right.

Handle fitted

The making of sword furniture could have been in the hands of goldsmiths for prestige weapons, however, the handles from the Nydam swords are quite plain and could have been made by semi-skilled wood turners. When considering all the skills that are necessary to produce a single sword it seems unlikely that one man will have all the relevant skills at his command.

If the model proposed above is followed we can see that although the number of hours does not diminish, the type of labour involved has gone from highly skilled to semi-skilled for a large proportion of the production cycle. Indeed, the most time-consuming portion, that of forging the strips, can move into the hands of semi-skilled and perhaps even unskilled labour. This would bring the cost down. A precedent for this model can be seen in the manufacture of the Samurai sword in Japan (Kapp et al. 1987). Closer to home in Sheffield the blade manufactures had blacksmiths who forged blade blanks (known as "little mesters" (little masters)). These blanks then went on to the cutlers who turned them into finished articles.

Discussion

The pattern-welded sword made in this experiment involved about two weeks work. A blacksmith working alone will only be able to produce 25 swords in one year. Labour has to be paid for, and it is paid for either in cash or in kind, depending on the status of the craftsman, i.e. free man or slave. The sword produced in this experiment was simple. Swords with six or eight bundles, as well as the cutting edges, are not unknown. Such a sword could well take twice or even three times as long to make; this would mean that it would be a very expensive sword indeed.

The level of expertise demanded in producing a pattern-welded sword is high, but should be within the level of skills of any competent blacksmith. However, the skills shown in the swords examined by the author show that the skills exercised were very high, indicating the sword-smith could well have been a specialist in pattern-welding. A possible alternative is that the smith carried out only the welding. Pre-forging of iron into strips or rods and the subsequent finishing might be performed by other craftsmen.

Conclusion

The experiments conducted leave very little doubt that the manufacture of swords was carried out by a number of different craftsmen with varying degrees of skill. Pattern-welding was used for both decorative purposes and to improve the mechanical properties of the metal. The pattern-welded sword was an expensive item compared with other iron artifacts.

4.8 Chain-mail: Experiments to reproduce the techniques of manufacture

Protective armour made from interlocking rings of metal (*lorica hamata*) was developed by the Celts and later adopted by the Roman army Bishop and Coulston (1993). This type of armour is often referred to as chain-mail. It afforded adequate protection against cuts from edged weapons such as swords, and thrusts from wide-bladed spears. Although it is known to have existed in the Republican period, the extent of its at that time is uncertain. The subsequent use of this sort of armour throughout the Roman period and beyond shows that it was popular with those who had to rely on it for protection.

Previous experiments by the author have shown that the making of chain-mail coats is a time-consuming undertaking. The making of a single mail coat is a daunting task and it follows that manufacturing mail on even a modest scale represents a considerable investment in material and, more importantly, time.

The manufacturing techniques used to make chain mail in the Roman period have been studied and a possible method using sharpened punches was suggested by Biek (1963). Experimental study of the possible methods of manufacture may lead to a better understanding of this area of Roman technology.

There are several methods of constructing mail coats, such as; using only butted rings with no rivets, or with all the rings riveted. The method of coat construction studied here is a combination of a solid ring with four riveted rings through it (Fig. 4.17). Mail coats made to this configuration are stronger than those made from rings that are butted together. The riveted rings are made from wire; the solid rings have no joints or welds.

This section discusses experimental methods that were undertaken to investigate possible methods of manufacture of both wire and, solid rings. The first part of this section will consider the making of the solid rings. While the fabrication of the riveted rings is clearly understood, this is not so for forming the wire from which they are made. This issue will be considered in the second part of this section.

This type of armour was worn through out the Roman period by legionaries and auxiliaries, both infantry and cavalry. Very little survives from the Republican period but there are representational examples from the 1st century B.C. (Robinson 1975).

The styles of coats and methods of ring construction have many variations. These may be related to the quality of the finished product, and it is possible some coats were designed to fulfil a definite function, such as those that gave protection to the body but not to the arms. It was not the object of this set of experiments to address these specific questions.

Consideration of solid rings

Rings that have no obvious method of jointing, such as welds or rivets, will be referred to as "solid rings". Such rings could easily be produced today using modern punch tools, whose manufacture requires precision machinery which is thought not to have existed in the Roman period. It is naturally assumed that precision punching was not the method used to produce solid rings. However a high degree of precision can be achieved without the use of high precision machines. A set of experiments was conducted to show that a punch and die can be made to a high degree of accuracy using only unsophisticated tools.

Examination of original rings

The analysis of Roman chain-mail rings of iron presents many difficulties due to their generally poor condition and, the very small number in existence. It is also very difficult in some cases to be certain if the rings are of Roman manufacture or were made outside the empire but copying Roman patterns. In many cases the amount of information that can be determined is limited because the rings are fused together with iron oxide and, when examined, are often found to consist of a thin shell of iron with a hollow centre. Such is the case with the rings found at Lower Bridge St, Chester in 1993, now in The Grosvenor Museum (small finds No 242.252 and 257).

Also examined were iron rings from Thorsberg and Nydam (Engelhard 1866), Raddaz (1968) has proposed a date of between A.D. 150-250 for these rings. As these rings were found outside the Empire it is not possible to say if they are of Roman manufacture or were made by barbarian tribes copying Roman originals.

Many mail rings from the Roman period are fused together into solid lumps and it is not possible to detach single rings for examination. Some individual rings exist and were made available to the author for examination. Twenty fine guided bronze rings from Leiden, that were found in 1902 in a town called Ouddorp, on the island of Goedereede, Provence of South Holland and are illustrated by Robinson (1975). The context is unclear but the armour was found beneath a layer of clay under some Roman pottery. Iron rings from Caerleon (found in Via Principalls 207 phase 5B , second century) were also examined.

Non-ferrous rings

The non-ferrous rings from Leiden, Holland, were assembled so that a single riveted ring had four solid rings inside it (Fig. 4.17). Six solid rings were examined under a microscope and measured on a shadow graph (see Table 4.9).

Ring No	Position	O/D	I/D	Thickness
1	1	3.110	2.212	0.51
	2	3.147	2.123	0.58
2	1	3.057	2.209	0.61
	2	3.013	2.257	0.63
3	1	3.174	2.203	0.48
	2	3.062	2.212	0.52
4	1	3.147	2.129	0.54
	2	3.133	2.209	0.59
5	1	3.054	2.268	0.57
	2	3.100	2.257	0.57
6	1	3.197	2.213	0.61
	2	3.196	2.234	0.58

Table 4.9 Non-ferrous solid rings from Leiden, Holland

The outside diameters of the solid rings measure 3 mm. When examined under a microscope the shape of the solid rings showed that these rings were probably made by punching from a solid sheet. If the technology was in existence to punch small rings from non-ferrous metal then it is possible that the same technology was used to punch rings from sheet iron. Some solid ferrous rings from Caerleon (found in Via Principall 207 phase 5B , second century), were examined under a microscope and measured on a shadow graph (see Table 4.9).

Ring No	Position	O/D mm	I/D mm	Thickness mm
1	1	6.957	4.968	0.989
	2	6.99	4.869	0.948
2	1	6.761	4.865	0.852
	2	6.948	4.765	1.513
3	1	7.381	4.791	1.263
	2	7.344	4.909	1.177
4	1	7.401	5.169	1.119
	2	7.513	5.465	1.116
5	1	6.703	5.024	0.864
	2	6.945	5.006	0.929
6	1	7.087	4.827	1.071
	2	7.247	5.042	1.036
7	1	7.605	4.818	1.501
	2	7.503	4.761	1.431
8	1	6.645	4.413	1.126
	2	6.609	4.356	1.071
9	1	7.850	4.939	1.393
	2	7.736	4.820	1.441
10	1	6.871	4.567	1.239
	2	6.974	4.465	1.441

Table 4.10 Solid ring from Caerleon

Reproduction of solid rings

Table 4.10 shows that the variation in the dimension of the outside diameter of the Caerleon rings is very small - of order of ± 0.6 mm as is the inside diameter. The consistency of these dimensions is all the more surprising, when the factors of wear and corrosion are considered. This suggests that the rings were made to a very high degree of accuracy/reproducibility. It follows from this, that Roman tools could be made to a high degree of accuracy. There are various methods which might be employed to manufacture the solid rings, for example:

(a) A solid bar has its centre removed by drilling, and the resulting tube is cut into thin rings.

(b) A sheet of metal has a hole drilled in it, this is the inside diameter of the ring and the outside diameter is formed by filing.

(c) Rings are produced by pearcing the inside diameter and punching the outside diameter when the metal is hot.

(d) Use of punch/die set

All the above were made the subject of experimentation.

Experiment (a) A solid bar was drilled with a spade drill driven by a bow drill. After 45 minutes of drilling the drill had made a penetration of 1.0 mm, this was not enough to make a single ring. The time taken showed that this method was not a feasable method to make solid rings.

Experiment (b) A sheet of wrought iron 1.5 mm thick was drilled using a spade bit in a bow drill. It took 20 minutes to drill a 6 mm diameter hole. The sheet was cut so that the hole was in the centre of a square piece of metal. This was then filed into a anulus. This took a further 12 minutes, giving a production time of 32 minutes for each ring. This too is an unreasonable method of ring production.

Experiment (c) A strip of wrought iron was heated and a hole punched through it to form the inside diameter of the ring. A punch had been made to cut the outside diameter of the ring. Although punching the inside diameter took only a few seconds the punch that had taken 45 minutes to make, only cut four rings before it was blunted beyond repair.

None of the above methods offered a realistic method of producing solid rings. Preliminary experiments showed that punching an annulus with one strike is very difficult with simple tools, because the geometry of the punch tool needed makes it too weak to withstand the forces involved in shearing the iron strip. Rings can be produced by using two sets of punches and dies. The first set punches the inside diameter of the ring and the second set the outside diameter. The use of more than one set of punches could account for any variation in the outside diameter of the rings.

A set of experiments was conducted to investigate whether punch tools of high consistency could be produced by the technology available in the Roman period, which did not require the use of precision machinery. The time taken to produce solid rings was recorded. The manufactured rings were measured on a shadow graph and the results are presented in table 4.11.

The solid rings were made as follows: In order to make the punch and die to punch the inside diameter of the ring, a piece of medium carbon steel was forged into a rod using a pair of swages as shown in Fig. 4.18 a. One end of the bar was then forged into a tapering point and the bar cut in half. The tapered punch 6.0 mm in diameter (A in Fig. 4.18 a). was used to punch a hole through a block of iron 25 mm square that had been heated to 950°C (Fig. 4.18 b). The punch was quite badly damaged in this operation but was quickly repaired. On cooling, the diameter of the hole shrank by 0.074 mm which meant that the unused piece of bar (B in Fig. 4.18 a) was too large to fit in the hole. The bar was hardened and tempered.

The bar had to be made to fit the hole with the minimum of clearance between the two. This was done by using a mixture of finely powdered glass mixed with water as a grinding agent. The mixture was smeared on the punch, the bar was then put over the hole and a bow drills used to impart a reciprocating motion to the punch (Fig.4.18 c). This caused the punch and die to be ground to fit each other, as the punch was hardened it is likely that most of the metal removed came from the die.

Great care was taken to keep the punch vertical. A smooth sliding fit between die and punch was achieved in 12 minutes. A slot was then cut into the end of the bar as shown in Fig.4.18 d. The slot was made 0.5 mm wider than the strip of iron that had been used to make the ring This was to allow clearance between the slot and the strip. Fig.4.18 e shows the punch and die with the strip in place. The strip was punched and produced a hole that would become the inside diameter of the ring.

In order to make the punch and die to punch the outside diameter of the ring a pair of swages was used to produce a bar with two separate but concentric diameters, the larger equal to the outside diameter of the ring and the smaller to the inside size as the holes in the strip (Fig. 4.18 f). The form of this punch is a tenon. (Tenons are made by blacksmiths as a method of fixing two pieces of iron together Fig. 4.18 g) . The die was heated to 950°C and a hole was drifted with the same punch as was used in the last die (Fig. 4.18 a). The stepped punch was then used to make a stepped hole as shown in (Fig. 4.18 h).

The punch was lapped to fit the die by the use of a bow drill with finely powdered glass as an abrasive. Figure 4.18 i shows the punch and die being used. The smaller diameter of the punch located the hole in the strip and then was located in the small hole in the die (Fig. 4.18 j). The punch was then struck with a hammer and the ring cut out. This system was used to produce thirty rings.

The Caerleon rings have no remaining external markings. This would probably have worn off with use of the coat, so it is impossible to determine if any further work was carried out after punching. Microscopic examination of the rings from Thorsberg and Nydam shows small flats on the outside diameter of the ring whilst the inside is very smooth. It is possible that the ring had been hammered on a mandrill. To test this theory, two of the rings from the previous experiment were placed on a tapering mandrill and hammered (Fig. 4.18 k). This had the effect of thinning the ring thickness and increasing the width. The two rings were examined side by side with the originals and the hammer marks on the surface were almost identical. It is likely that rings such as these were produced by punching, followed by hammering on a mandrill. Experiments showed that as the punch and die start to wear the rings become somewhat distorted. The distortions are easily removed when the ring is hammered on a mandrill.

Ring No	Position	O/D mm	I/D mm	Thickness mm
1	1	9.491	6.371	1.52
	2	9.000	6.402	1.59
2	1	9.557	6.36	1.51
	2	9.679	6.54	1.56
3	1	9.419	6.231	1.48
	2	9.322	6.474	1.49
4	1	9.750	6.294	1.52
	2	9.623	6.316	1.58
5	1	9.560	6.495	1.59
	2	9.625	6.288	1.61
6	1	9.524	6.315	1.46
	2	9.819	6.603	1.49
7	1	9.550	6.533	1.59
	2	9.389	6.277	1.49
8	1	9.464	6.291	1.47
	2	9.505	6.365	1.41
9	1	9.698	6.544	1.54
	2	9.551	6.382	1.55
10	1	9.610	6.405	1.51
		9.629	6.647	1.69

Table 4.11 Experimental Rings.

In further tests, Vickers Hardness measurements were carried out on the rings from Thorsberg and Nydam and were compared with some of the experimentally produced rings. The results are shown in table 4.12.

It can be seen that the experimental ring which was hammered on a mandrill after punching is harder than the ring that has only been punched, indicating that some work hardening has taken place. The experimental punched ring had a hardness value of 146 VPN, which is very close to the 149 VPN of the metal from sample A. This indicates that punching did not produce any significant work hardening. The experimental ring that was hammered on a mandrill increased in hardness from 146 VPN to 210 VPN which is in the same region of hardness as the Roman originals. The hammer marks together with the hardness of the rings would suggests that work hardening was brought about by hammering. Thus hammering on a mandrill improves the hardness of the ring because of work hardening and also the appearance of poorly punched rings.

The specimens from Nydam and Thorsberg were polished and etched and then examined and photographed. A similar procedure was carried out on two rings that had been produced experimentally. The first ring had only been punched and, the inside edge shows the deformation associated with punching. The second ring had been punched and then hammered on a tapering mandrill. This had distorted the grains and left little evidence of the previous punching. When the latter was compared with the ancient examples it was found that there are some similarities in the structure. To produce one solid ring by punching took 21 seconds, while hammering it to shape on a taper mandrill took 50 seconds. With further practice the author was able to reduce this time to 32 seconds. As with any skill, practice increases speed and the time to punch and hammer the ring on a mandrill was reduced to one minute.

Sample	Vickers Hardness Number
A (hammer and anvil)	149
A (swaged)	138
Thorsberg ring	191
Nydam ring	187
Experimental ring punched then hammered on a mandrill	187
Experimental ring (punched)	210

Table 4.12 Vickers hardness tests on solid rings.

Conclusions

The experiments to make punches and dies using simple technology have shown that such tools can be made to a high degree of precision using the level of technology that would have been available in the Roman period. Such tools could be produced quickly and maintenance was quite simple. When a die wore out, a new die could have been made in less than twenty minutes. The comparison of the sizes of the ancient Roman rings with experimental rings shows that the size ranges/variations are within the same order as those found in the ancient rings. Thus, they and could have been made using punches and dies similar to those made experimentally.

The Roman rings that have hammer marks have a hardness that is almost the same as that of experimental rings made by punching and hammering on a mandrill. The hardness of the ancient rings is higher than the hardness of pieces of wrought iron that have been forged and normalised, indicating that work-hardening has taken place. It is suggested that this hardening is produced by hammering the outside of the ring and, as the rings are circular, it is probable that the hammering was carried out on a mandrill (possibly tapering). A ring can be punched from a sheet and hammered in to shape in 60 seconds.

The manufacture of wire for the production of riveted rings

Solid rings cannot be joined together unless a joining ring can be opened to fit the solid rings in it. The joining rings can be made by several methods such as butting the ends of the rings together, welding them together, or joining them by riveting. The manufacture of welded rings and butted rings is still the subject of investigation, so only riveted rings will be considered here. To make a riveted ring, wire is prepared and then wound round a cylinder to create a coil. The coil is cut down its length to give a series of rings. The ends of the rings are flattened and a hole punched in each flat. The rings are closed by putting a rivet through the hole.

Experiments were conducted to determine the possible methods that could be used to produce the wire to make riveted rings. Very little is known about the wire used to make chain mail because there are so few samples to examine. However a certain amount of knowledge can be gained from work on other artifacts that can be applied to wire. Steel and wrought iron are the only ferrous metals used in the Roman world. Steel is thought to have been too expensive to manufacture chain mail so the only other metal left is wrought iron (the iron produced from the bloomery process). Experiments by the author have shown that as the cross-sectional area of the metal decreases, splitting and cracking will occur unless the iron has had most of its slag reduced to such a size that they will not assist in forming cracks (Sim 1994). So we can infer that the material used for making chain mail was good quality wrought iron. Wire drawing is performed by pulling a piece of metal through a die plate in which there are series of holes that decrease by approximately 10% of the wire diameter per hole. The decrease in hole size can be less than 10%, but if it is larger, the force required to draw the wire is greater than the tensile strength of the metal and so it fractures.

To gain accurate dimensions for the reproduction of chain-mail, ancient rings were measured to obtain the size and shape of the wire and the diameter, and method of assembly of the rings was examined. Ferrous chain-mail rings from Caerleon, Wales, and also some from Thorsberg, Germany were examined. The rings were made from wire which was square or rectangular in section, rather than circular.

The tables below, (4.13 and 4.14) show that the wire for these rings was made to very tight tolerances. Different production methods were considered for the manufacture of the quality and quantity of iron wire needed for this application.

In the tables below, dimension A is the width of the ring, and B its thickness.

Ring No.	Position	Mean size (mm)	Deviation from mean (mm)
1	A	1.07	+ 0.03 – 0.07
	B	1.20	+ 0.00 –0.00
2	A	1.00	+ 0.00 – 0.00
	B	0.89	+ 0.01 – 0.01
3	A	0.90	+0.10 – 0.02
	B	1.32	+ 0.08 – 0.04
4	A	1.06	+ 0.02 – 0.04
	B	1.29	+ 0.01 – 0.01
5	A	1.20	+ 0.00 – 0.00
	B	0.94	+ 0.02 – 0.04

Table 4.13 Riveted rings from Caerleon

Ring No.	Position	Mean size (mm)	Deviation from mean (mm)
1	A	1.04	+ 0.36 – 0.02
	B	1.92	+ 0.08 – 0.12
2	A	1.45	+ 0.15 – 0.17
	B	0.15	+ 0.06 – 0.04
3	A	1.25	+ 0.15 – 0.14
	B	1.80	+ 0.00 – 0.00
4	A	1.57	+ 0.10 – 0.05
	B	0.02	+ 0.02 – 0.00

Table 4.14 Riveted rings from Thorsberg

Methods of producing the wire

The material for each ring can be considered to be wire, because of its dimensions (cross-sectional area). There are four ways in which this material can be produced.

(a) The wire is hammered down from a bloom into wire;

(b) The wire is formed between a pair of swages;

(c) The wire is produced by drawing through a draw plate;

(d) The wire is produced by passing it between grooved rollers.

Each of these methods will be considered in further detail below.

Hypothesis (a) It is possible to forge iron down into wire using only a hammer but it presents many difficulties. When the cross-section of the wire gets below 5.00 mm diameter, the metal cools very quickly, due to heat loss to the atmosphere and the anvil and hammer absorbing heat. It is also very difficult to handle, except when it is in very short lengths, of 80.00 mm or less. Placing it in the fire presents a major problem because it bends. The only way is to coil it, and by the time it is straightened out it has lost most of its heat. Even the most skilled smith would be very hard pressed to produce wire to the tolerances shown in these rings.

(b) Wire can be produced by forming it between a pair of swages (see glossary). Both top and bottom swages are known from Roman times, although they have not been found together. It is therefore possible to produce wire in this method. It would be necessary to pass the metal through pairs of swages that become progressively smaller until the finished size is reached. The problems are the same as a) for handling, and only short lengths could be produced.

(c) The most sensible way of producing wire is to draw it through a series of tapering holes that diminish in size until the required size is obtained. According to Singer et al.(1954) draw-plates for drawing wire were unknown until the tenth century AD and wire was first produced by forging. However Thomsen and Thomsen (1976) have stated that non-ferrous wire from Persia of 6th-5th century B.C. date could only have been made by wire drawing. Northover (1995) has argued that two bronze plates found at Iselham, Cambridgeshire, thought to belong to the Late Bronze age, are draw-plates. These draw-plates are assumed to be for drawing non-ferrous metal. The drawing of iron wire is more difficult because the higher tensile strength of non-ferrous metals requires a much greater force to bring about drawing. The force could easily have been supplied by the winch system used for drawing back the firing arms on ballista (Marsden,1971).

Thomsen and Thomsen (1976) have shown that wire can be drawn through a plate made of the same metal as the wire to be drawn. They have drawn gold , silver, and copper wire through gold silver and copper draw plates. If this theory holds true for iron, then draw-plates for drawing iron would not have to have been made from hardened steel. In another paper Thomsen and Thomsen (1974) have shown that if the hole in the die is in the form of a taper, drawing will take place if the draw angle is >2° and <50°. A series of tapering holes can easily be made in a piece of iron by making a tapering punch and driving it into the plate to different depths. This would be done as a forging operation.

In a chain-mail coat there could have been as many as 12000 riveted rings if the ring diameters were in the region of 12 mm, and even more for smaller rings. Examination of Roman chain-mail rings has shown that their dimensional accuracy is very close. Experiments by the author show that making wire by hammering and filing is unlikely to be able to produce wire to the tolerances found in Roman chain-mail coats or in the necessary quantities.

A draw-plate was made to test the proposition that iron wire can be drawn successfully through an unhardened iron draw plate and that the punch for forming the holes was square tapering and made from plain carbon steel. The draw-plate, 4mm thick x 36mm x 120mm, was forged from a piece of wrought iron bar stock 20mm in thickness. A series of square holes of increasing size were punched with a square section punch (Fig. 4.19). All forging was conducted at forging heat (1000°C) and the iron was quenched after the final heat. There was no cold working. The times to make the punch and the draw-plate are given in Table 4.15.

Operation	Time (minutes)
Forging the punch	11
File to shape	10
Polish on sandstone	15
Forge wrought iron 40 mm x 20mm down to 4 mm x 36mm	12
Punch 6 holes	21
Total time	1 hour 19 minutes

Table 4.15 Sequence of operations, and time for production of an iron draw plate.

Before forging, the draw-plate had a hardness number of 108 on the Vickers scale (108 VPN) after forging it increased to 127 VPN This indicates that some work-hardening had taken place.

The first stage of preparing the wire for drawing was to forge a length of wrought iron down to 3.5mm square and then anneal it by heating to red heat and burying it in wood ash. After annealing the hardness was 100 VPN scale. The wire was lubricated by wiping it with beeswax and drawn through holes 1-5 inclusive on the draw-plate using an INSTRON 4206 Universal Testing Machine. The results are shown in table 4.16. No annealing was conducted between each drawing and due to work hardening of the finished wire, the hardness had further increased in hardness to 159 VPN which is harder than the holes in the draw-plate which was 127 VPN. It should be noted that because hardness values are not on a linear scale the relative hardness of the draw-plate to the annealed wire is be misleading.

Hole Number	Maximum Load (N)
1	1415
2	2094
3	2199
4	1146
5	1194

Table 4.16 Force required to draw wire through the drawplate

It can be seen that the force to draw the wire is substantial but it is possible that this could be supplied by man-power. However this would be working almost to the limits of human strength and it is uncertain how long an individual could keep on exerting this sort of force. It is possible for the Romans to have had a winding device, to provide the power to draw iron wire. In fact a small capstan, or windlass would have been sufficient to provide the necessary power to draw the wire through the plate

This simple experiment shows that iron wire can be drawn through a soft iron plate and that a draw-plate and the tools to make it can be produced in 1 hour 19 minutes. It is possible that some work-hardening will take place on the inside of the holes. Annealing of the wire in between drawings would reduce the power needed for drawing; as would using smaller taper angles.

Roman draw-plates from Germany and Britain

The use of draw-plates by the Romans has been established by Thomsen and Thomsen. who describe a draw-plate in the Burg Altena Museum in Altena, Germany. The draw-plate from Altena in Germany (Nr. Dusseldorf) whose dimensions are given in Table 4.17 exists as a cast, the original having been lost. It has been dated by the museum to c A.D. 45 and is thought to be of Native/Roman origin, but no more details are available. The similarities with a second draw-plate from Vindolanda are apparent not only in form but also in size. The main difference is that the Altena plate has a groove down the centre. The purpose of this is not certain but it could have been for holding a lubricant such as wax or animal fat.

An iron object found at the Roman fort of Vindolanda is shown in Fig. 4.20. The general appearance of this object indicates a well made tool. There are no hammer marks visible and care has been taken to produce fairly flat and smooth surfaces. There is a small amount of corrosion inside some of the holes, but otherwise the find is in excellent condition.

The plate is pierced by four holes and, at the broken end, there are the remains of a fifth hole. Inspection revealed that the holes taper. They are not perfectly circular but the deviation from true circularity is so small as to be negligible. The holes are fairly evenly spaced along the centre line of the plate.

Examination with a microscope at magnification X 60 shows wear marks inside the smaller ends of the holes, although in two holes (diameters respectively 2.8 mm and 3.2 mm) some of the wear marks have been obliterated by corrosion. Until an X-ray of the draw plate is available it is impossible to describe with absolute certainty the exact form of the holes.

Method of manufacture of plate

From inspection of the side of the plate it can be seen that it has been made of at least three pieces of metal, pile-welded together. Considering the difficulties of drilling iron using a hand-powered drill, the holes were most likely to have been made by punching. The large holes on one side of the die are almost the same size and were most probably made with the same punch. This punch would have been tapered in order to punch holes to within 2 mm to 3 mm of the front face of the draw-plate. The smaller holes would have been made with a small punch and then the ends of the holes could have been smoothed with a file. The taper in the holes provided the angle necessary to draw wire.

Fig. 4.20 is a schematic representation of the two draw plates, the letters A to R indicate the dimensions. The dimensions are shown in the table below and the sizes of the two plates are compared in Table 4.17.

Dimension	Vindolada Size of dimension (mm)	Altena Size of dimension (mm)
A	6.20	6.90
B	6.20	6.62
C	6.30	6.02
D	6.30	4.94
E	3.20	4.78
F	2.80	3.82
G	2.60	3.23
H	2.30	3.14
J	26.00	38.80
K	20.00	25.00
L	13.70	19.00
M	93.00	105.00
N	25.00	19.60
P	19.80	17.25
Q	18.50	22.26
R	16.80	27.81

Table 4.17 Comparison of hole sizes in draw-plates from Vindolanda and Altena

We have seen in order to function as a draw-plate, it is necessary to have a piece of metal with a tapering hole where the angle of the taper is greater than 2° but less than 50.° The draw-plate from Vindolanda has a series of through, tapering holes that are positioned in ascending order of size, and it thus fits the criteria of a wire-drawn plate as described above.

Discussion

The size of the smallest rings (3 mm diameter) and the intricacy of the work involved in punching the holes for riveting puts this sort of work outside the normal work conducted by blacksmiths or armours. The use of mandrills and tiny punches is more the type of work conducted by a jeweller. It is not being suggested that the making of chain-mail was carried out by jewellers but that the equipment and the nature of the work would have employed the skills and tools of a jeweller rather than those of a blacksmith.

If we consider a mail coat constructed from a series of solid rings, with four riveted rings passing through it where the rings are 6 mm in diameter, then the whole coat, will require 170,353 rings (42,588 riveted rings and 127,764 solid rings). If each solid ring takes 1 minute to make then

1 (min) x 127,764 (rings) = 2129 hours

If it takes 3 minutes 47seconds to make a riveted ring (this includes making the rivet.

42,588 (riveted rings) x 3 minutes 47 seconds (per ring)=2684 hours 41 minutes

Total time 2129 hours + 2684 hours = **4813 hours** (1.82 years)

It is impossible to estimate how many coats were in use at any one time. Not all soldiers in a legion wore the same type of armour. But if one considers a single legion at full strength, with every legionary wearing the same armour then, if every man has one coat and there is one coat per man in store to act as backup then it would take;

6,000 x 4813 hours 41 mins = 28,878,000 man hours of work.

This is a staggering amount of time to invest in armour. It is possible that mail coats had a long life and it is known from some medieval mail that they were frequently repaired. But we are not sure how long a coat lasted or how much of the army were using mail at any one time. It is possible that there are other methods of manufacture we have no knowledge of, that might reduce the production time.

Certainly the use of solid rings reduces the time to make a coat. If the coat described above had been made of all riveted rings then it would have taken:

170,353 rings x 3 minutes 47 seconds (per riveted ring) =10,458 hours

10,455 hours as opposed to 4,813 hours 41 minutes a difference of 5,644 hours per coat. One possible reason for making a coat that is constructed from one riveted to every four solid rings is the amount of time saved.

Discussion

The measurement of ancient Roman solid rings has shown that they were able to produce rings to a high degree of accuracy, greater than that expected from hand methods such as forging. It seems likely that a punch was used to make these rings. Experiments have shown that:

(1) Simple technology can be used to produce a punch and die that are capable of punching rings from sheet iron.

(2) That the die and punches can be made in 45-50 minutes.

(3) That wear on the die can easily be rectified by putting the punch in the die and striking a single blow.

(4) The die can be made from soft iron but the punch needs to be made from steel or case-hardened wrought iron.

When experimental rings were compared with originals the amount of deviation from the mean was similar in both cases. It is suggested that solid Roman rings were made by punching with a punch and die similar to that used in the experiments. These punching sets were easy to produce and required no precision machinery.

Two iron plates from the Roman period have been shown to exhibit all the characteristics of draw plates, but it cannot be proved that they were used to draw iron wire. However, experiments have shown that iron wire can be drawn through iron draw-plates and that such draw-plates are simple to produce. During experiments the force required to draw iron wire was between 1415 N and 2094 N exerting this force by hand reaches the limits of human strength, but a simple system of pulleys would have been able to produce the necessary force.

The dimensional accuracy of the wire used to make Roman riveted rings was to a level that the author could not reproduce by forging and filing or by swaging. It is suggested that the accuracy of the rings and the time to produce the necessary quantities of iron wire for riveted rings means that the wire was probably made by drawing wire through draw-plates. The apparent scarcity of equipment for drawing iron wire can perhaps be attributed to some draw-plates having been misidentified as nail-heading tools, as well as the fragile nature of iron items in the archaeological record.

The amount of time to make chain mail by the experimental methods described in this chapter leads the author to believe that methods that could produce mail in a shorter time were in use during the Roman period. Investigation of other methods is currently in progress.

Conclusion

The mail coats examined display a high degree of dimensional accuracy in the individual rings. Experiments to produce rings by hand have failed to achieve the same degree of accuracy. It is suggested that in the case of solid rings, they were produced by punching from strips of iron using hardened punch tools. The difficulty of performing this operation by hand indicates that a press tool was used. The rings made from riveted wire also show a high degree of accuracy in overall dimensions and cross-sectional dimensions. Producing wire to this degree of accuracy was not possible using traditional hand forging methods; this indicates that either a forming tool such as a pair of swages was used, or the wire was produced by drawing. In view of the number of mail coats produced and the considerable amount of time required to make wire by hand-forging, there is a strong possibility that the wire was produced by drawing and the drawing was assisted by some mechanical means. Irrespective of the place of manufacture, the operations to produce the individual rings are not as complicated as would first appear; and it is possible to train a unskilled workforce, including children, to make the rings and assemble the coats. This is probable when one considers the time involved. A skilled craftsman would be wasted in making such a coat, when his services were necessary to produce items that could only be made by a skilled man.

4.9 Discussion

The experiments have demonstrated that even the manufacture of simple iron items represents a considerable amount of time and resources, first in preparing iron to make it suitable for forging, and turning that iron into artifacts. Second, a considerable amount of time has to be spent in bringing some artifacts to a finished state (the sword) while others can be used in the finished forged state (nails).

As has already been stated, wrought iron was the principal ferrous metal used in Roman times, and as with any material, it has both advantages and disadvantages. One of the disadvantages is the tendency to crack during forging. This is because of the slag inclusions present in all wrought iron. The effect of cracking can be reduced by refining by forging (see chapter 3). For most applications wrought iron has to be refined to a high standard for successful forging to take place. This is well demonstrated in the manufacture of the stylus. On first glance this is a simple item to make, and, as they were probably made in such large quantities, it is not unreasonable to assume that they could be made from metal of a mediocre quality. However the experiments show that items of small cross-section need high quality iron to start with or cracks will develop which will cause the forging to be discarded before it is completed. The same cracking effect was seen when wrought iron was forged into sheet to make some of the disposable weapons.

The wire for the chain mail coats also demonstrates the need for high quality in the basic product, because if the iron wire has large inclusions in it, the metal will break under the forces exerted when drawing takes place. The preparation of wrought iron in order to bring it to a high enough standard for forging is a long and expensive procedure and will add a significant cost to the price of the metal. Low quality wrought iron can be used for some applications where complicated or intricate forging is not required.

In our own era, such low quality iron was used to make railings for houses. No doubt there was a Roman equivalent and different grades of wrought iron had different applications. The items made for the experiments were complicated forging in terms of the techniques used to make them and, this type of work requires iron of a high standard.

Our modern perception of the use of hardened steel for artifacts is based on a plentiful supply of cheap steel. If steel is expensive this will restrict its use in weapons production. Alternatives will be found that are not so expensive. The hammer, stylus, and many of the disposable weapons show that there are definite advantages to not hardening some implements and this lack of hardening may have been a deliberate policy, and not due to lack of understanding. A hardened surface is only necessary where a particular item is subjected to continual wear, or when a tool is required to cut a hard material. So we see knives, like kitchen or craftsmen's knives, which are in constant use, need a cutting edge that is going to have a long life and will not need frequent sharpening. The same is true of woodworkers' chisels or woodsmen's axes. Frequent sharpening slows down production. If the owner can sharpen or repair a blade himself, it will be to his benefit, but a knife made of unhardened iron will wear away very quickly. It may be more economical to pay the extra cost of a steel knife because it gives a longer life span than two cheap knives. If a steel knife costs four times the price of a soft iron knife but lasts six times as long, it makes sense to buy a steel knife.

The use of steel tools which are left soft is demonstrated by the scythe and many other tools used in agriculture, where a soft-edged tool is a definite benefit.

"In pre-modern societies the making of steel was regarded as among the most difficult and complex of all industrial operations. Most of the direct process steels known to history were of poor or mediocre quality" (Rostoker, Bronson, and Dvorak, 1989, 11). Steel was known to be an expensive material in Roman times. A criticism of Roman technology is that it was inferior in the realm of iron-working, because the Romans did not harden and temper steel, or at least not very well. We have seen from the experiments that some items, such as the stylus, hammerhead, and many disposable weapons are more beneficial to their owner if they are not hardened, because the repair to them is very simple to carry out.

In the Roman period, the most common ferrous metal was wrought iron. Steel was available but very expensive, as is shown by its infrequent use and the high cost of turning wrought iron into steel by case hardening (Wagner 1990). Items made in large numbers will require storage and if not protected in this time will certainly rust. Theophilus (trans. Dodwell 1986) suggests various remedies to prevent rust and although it cannot be proved, it is reasonable to assume that some form of rust prevention was practised in Roman times. The use of iron was based on wrought iron and artifacts were made with this in mind. A technology based on a particular material will endeavour to take advantages of that material's strengths, and minimise its weaknesses. Therefore the advantages of tools and weapons deliberately made not to be hardened would be very clear to those working in the field and it is quite possible that items were left unhardened as a matter of deliberate working practice and not because of lack of understanding.

It has traditionally been assumed that the Roman iron industry was conducted on manpower alone and few if any mechanical devices were employed. If this is so, then the manpower invested in this industry was enormous, as can be seen from the times taken to produce the items made in the experiments described in this thesis. The notion that the army could be supplied from local smiths is not borne out by the experiments. Further, when large quantities of items are required, the most sensible method of production is for semi-skilled labour to be used as widely as possible, and for the skilled blacksmith to employ his talents on the work that his skills alone could fulfil.

The industry must have been well run and very efficient in order to produce the amount of iron which was made in the Roman period. Production costs must have been low but this is at odds with the man hours that experiments show are necessary to produce iron. If we assume that the experimental figures are correct and that the man hours are accurate, the notion that iron was cheap seems to be unfounded. The abandoning of such a large quantity of iron at Inchtuthill (Angus, Brown and Cleere 1962) can be taken to indicate that iron was a cheap metal; if this is so then the model for the production system needs to be modified. The form of this modification must be a matter for conjecture but one possibility is that not all the labour was manpower, and that some could have come from machinery. The cost of labour in man hours is large enough to be a significant factor in production costs of all iron artifacts made in the Roman period.

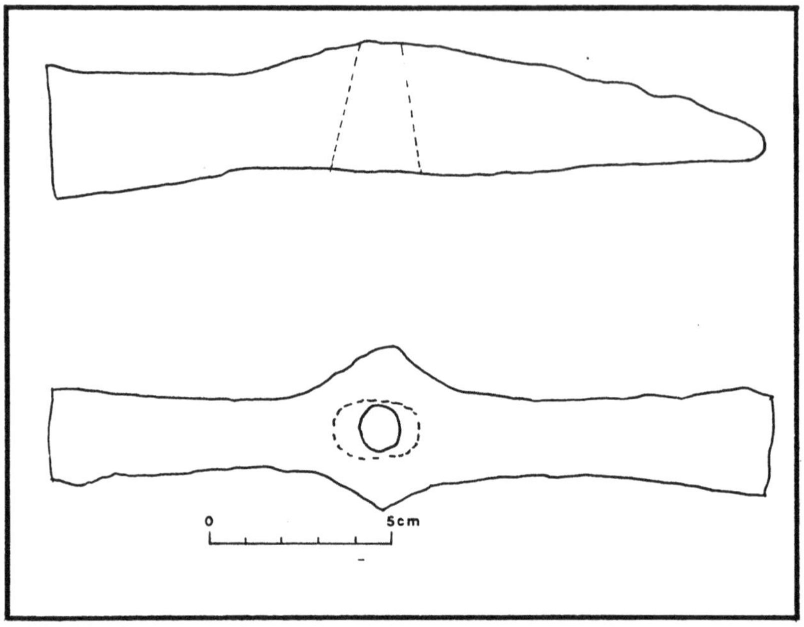

Figure 4.1 Outline drawing of a Roman hammer. (from the Silchester collection, Reading museum. ref 07510 yc 169)

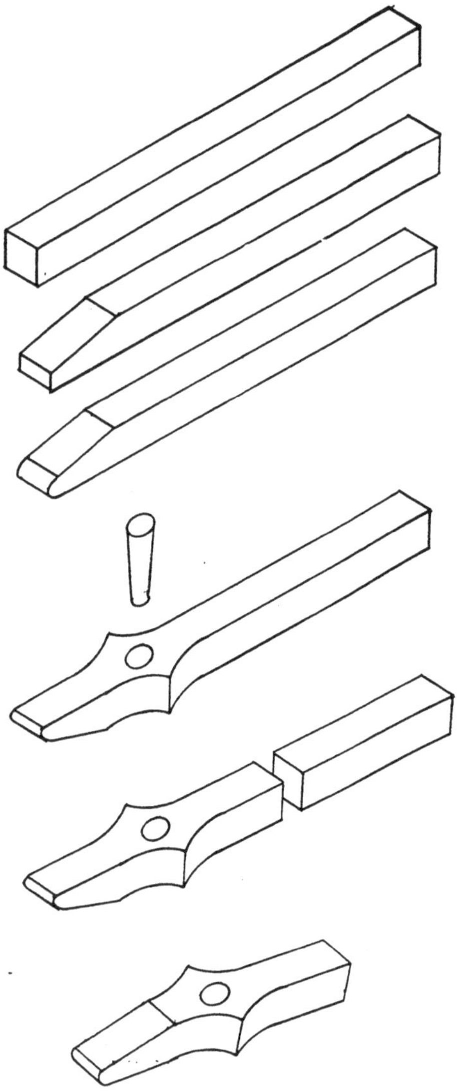

a Billet forged into bar.

b Peen end forged.

c Radius forged on the end.

d Eye-hole formed using a taper punch.

e Hammer head cut from bar.

f Striking end hammered flat.

Figure 4.2 Stages of manufacture of a hammer head

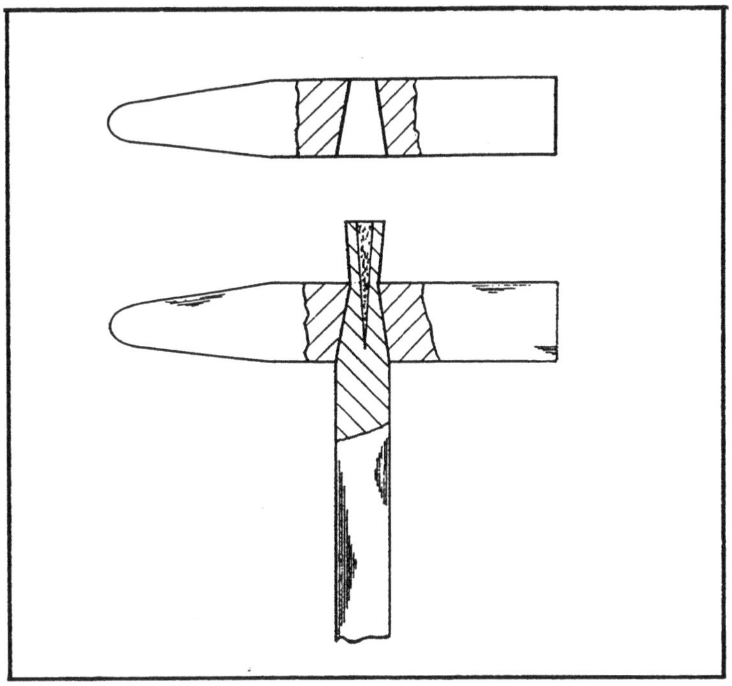

Figure 4.3 Section through a Roman hammer, showing the method of fixing the haft

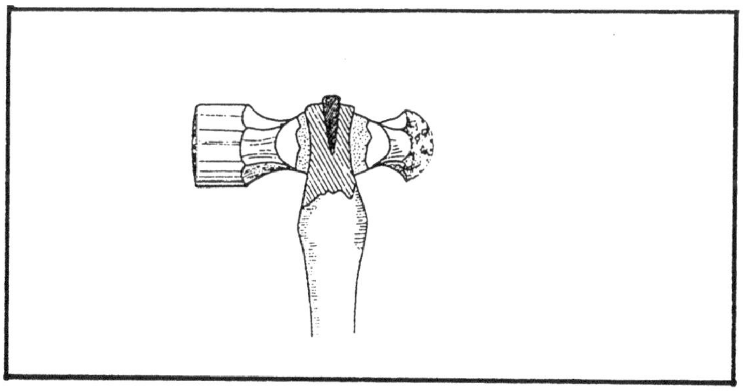

Figure 4.4 Section through a modern hammer

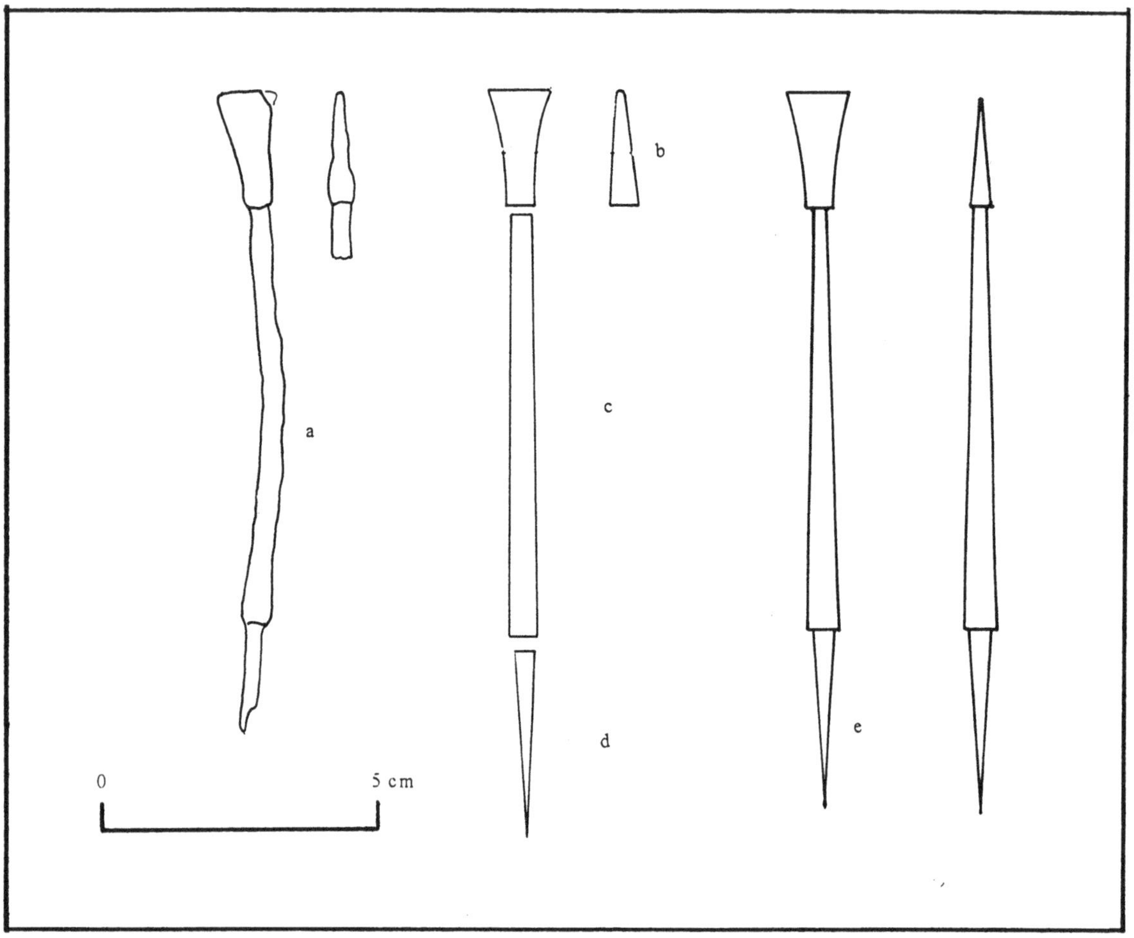

Figure 4.5 Drawing of Roman iron stylus

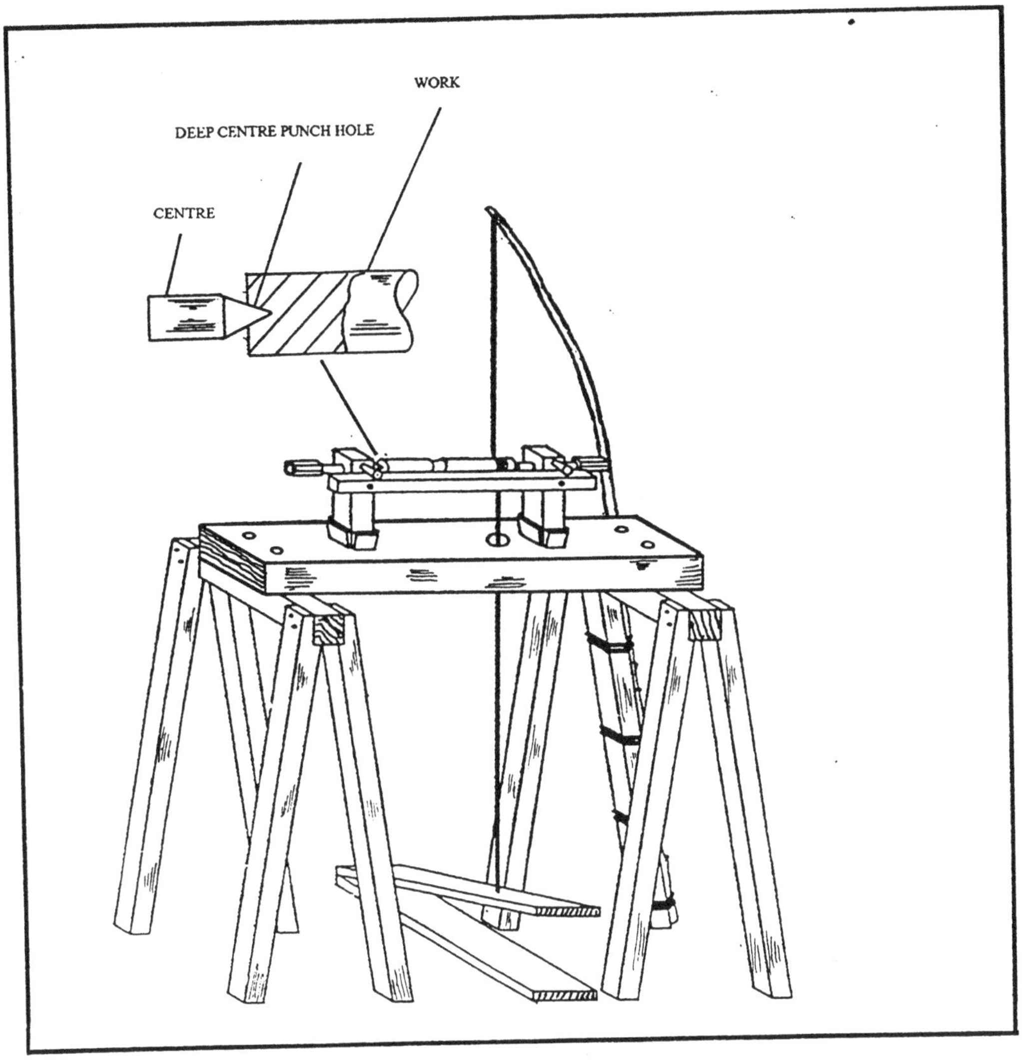

Figure 4.6 Pole lathe with enlarged view showing centres

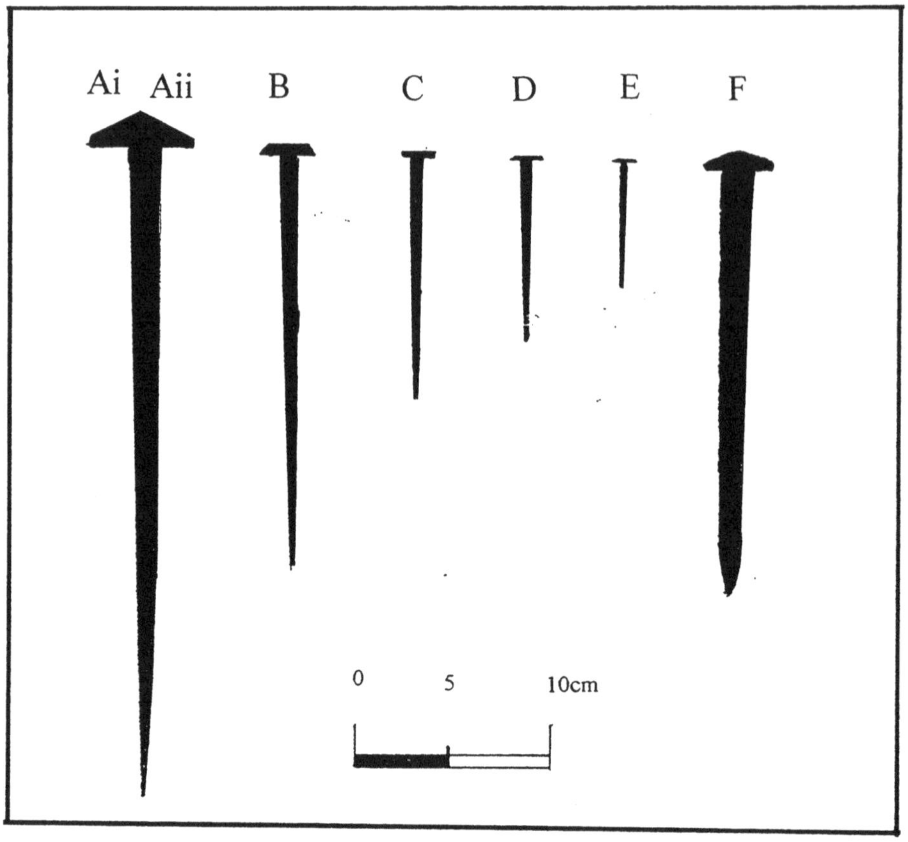

Figure 4.7 Nail type from the Inchtuthil hoard.(After Angus *et.al.* 1962)

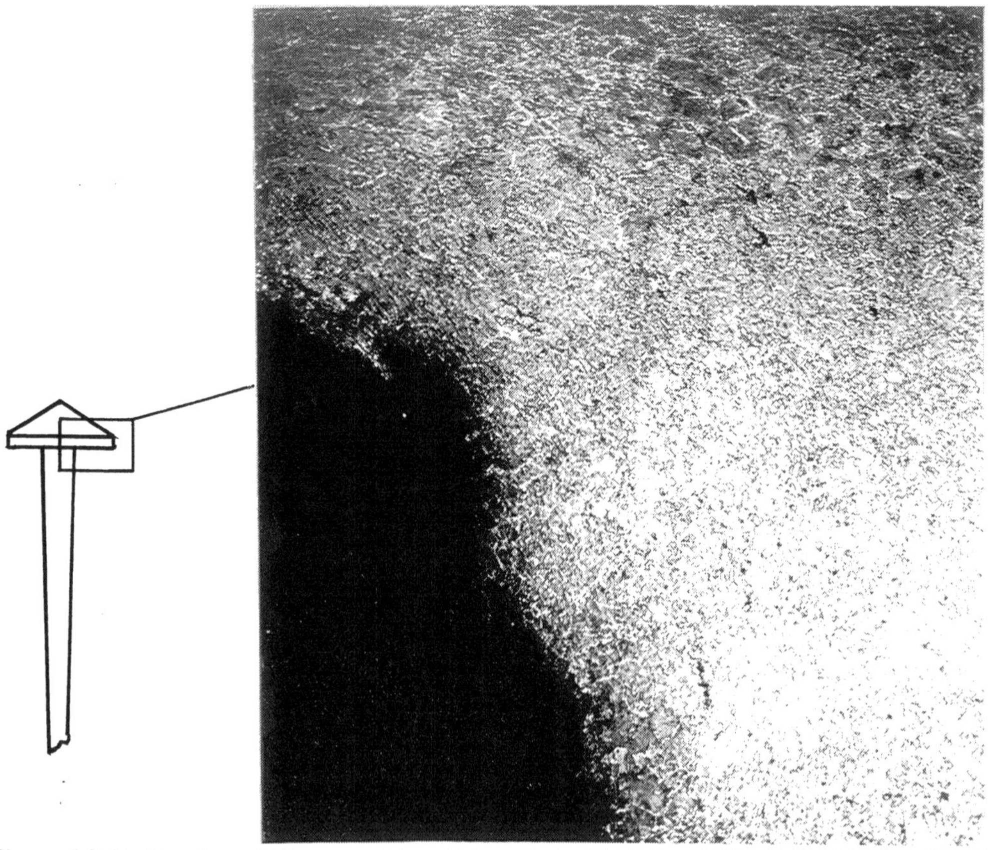

Figure 4.8 Section through Roman nail type D from the Inchtuthil hoard. Head of nail at top of picture. Magnification x 100

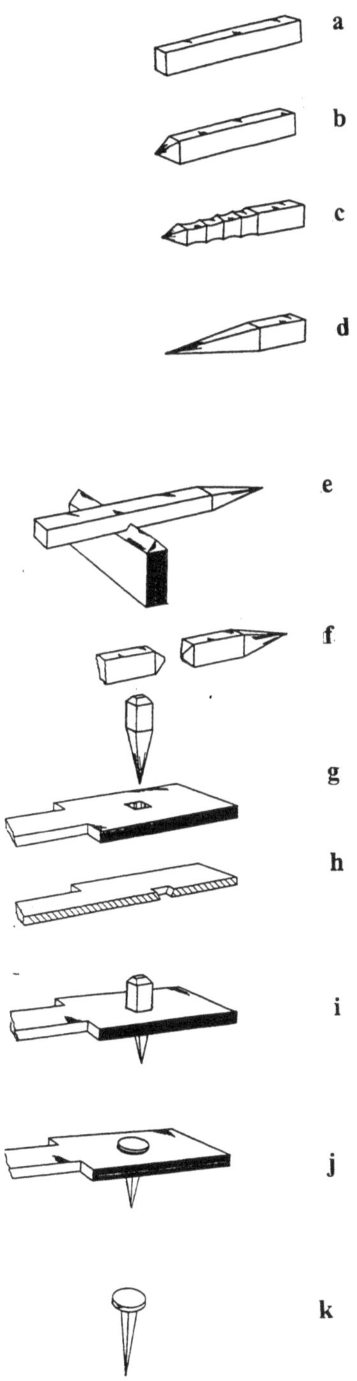

Figure 4.9 Stages of manufacture of a Roman nail

Figure 4.10 Scythe maker with three strikers (Amman 1989,70)

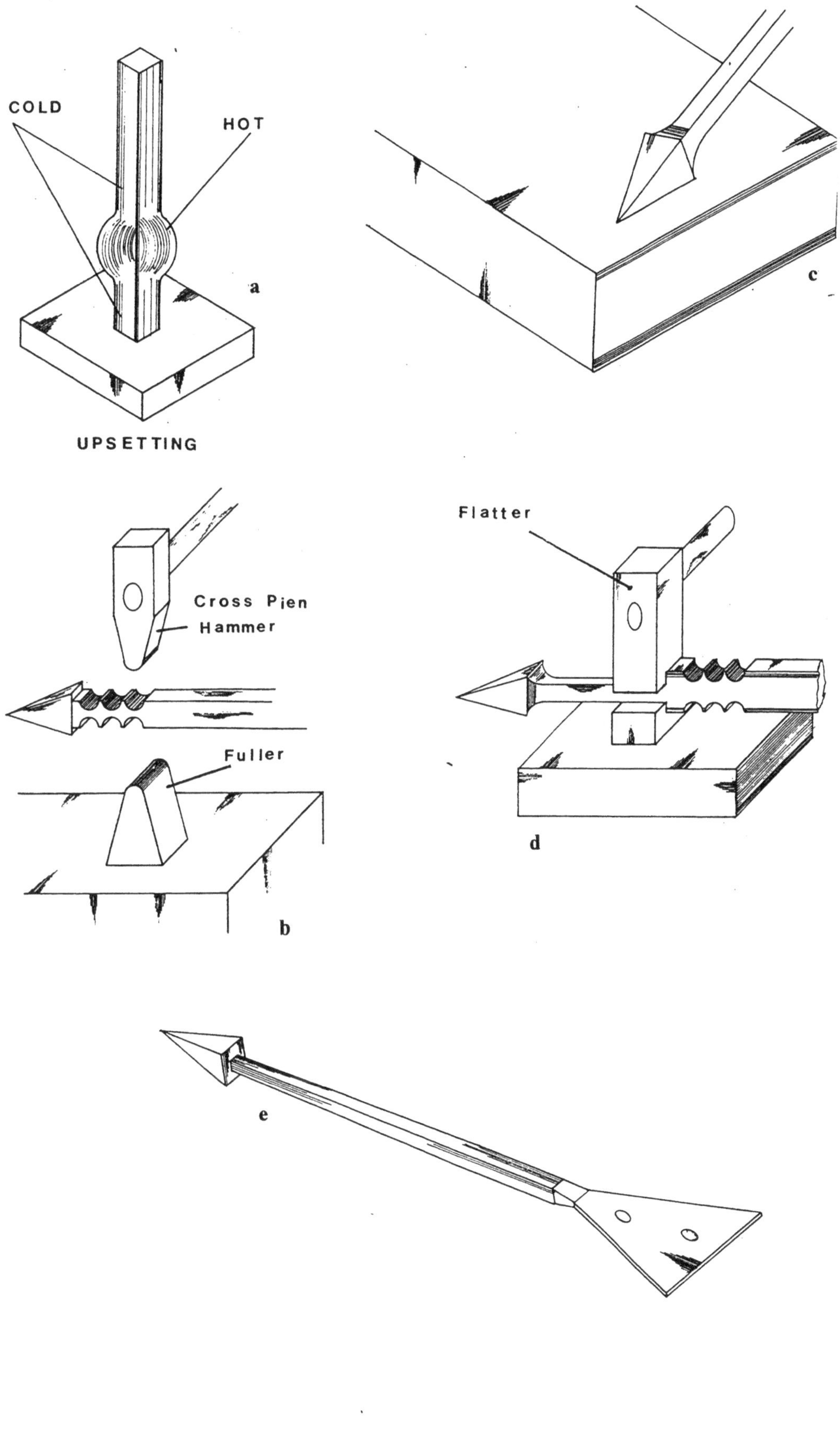

Figure 4. 11 Stages of manufacture of a Pilum

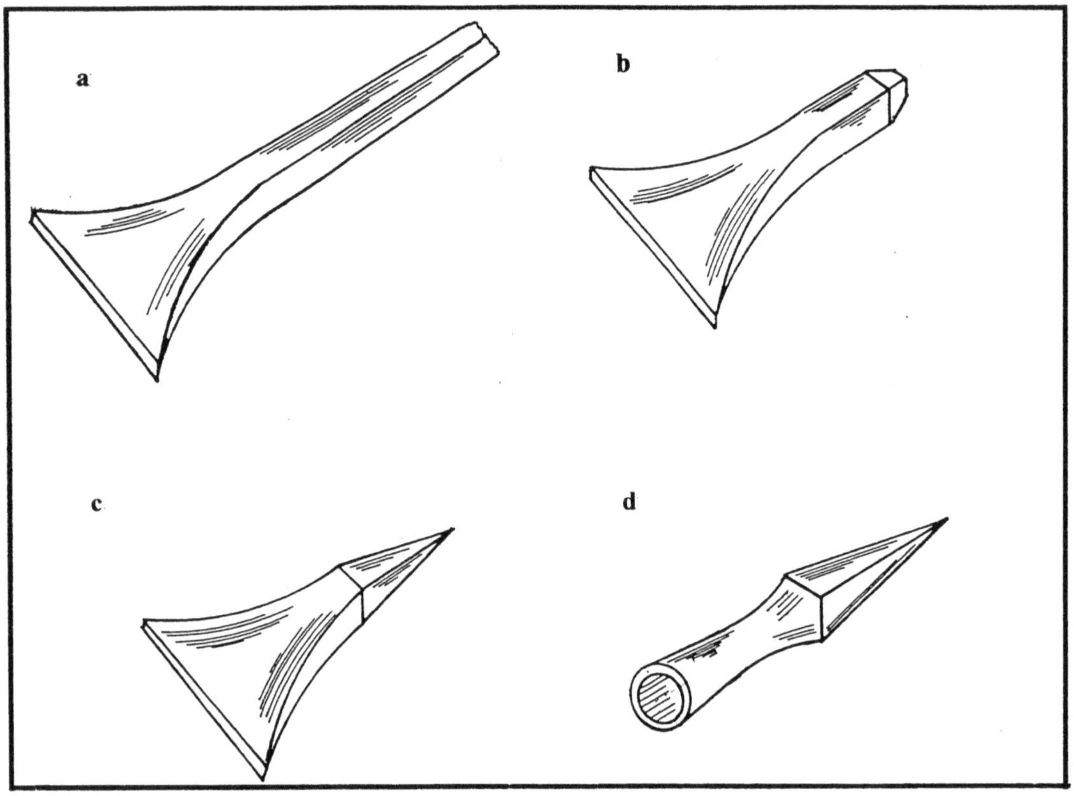

Figure 4.12 Stages of manufacture of a pyramid balista bolt head

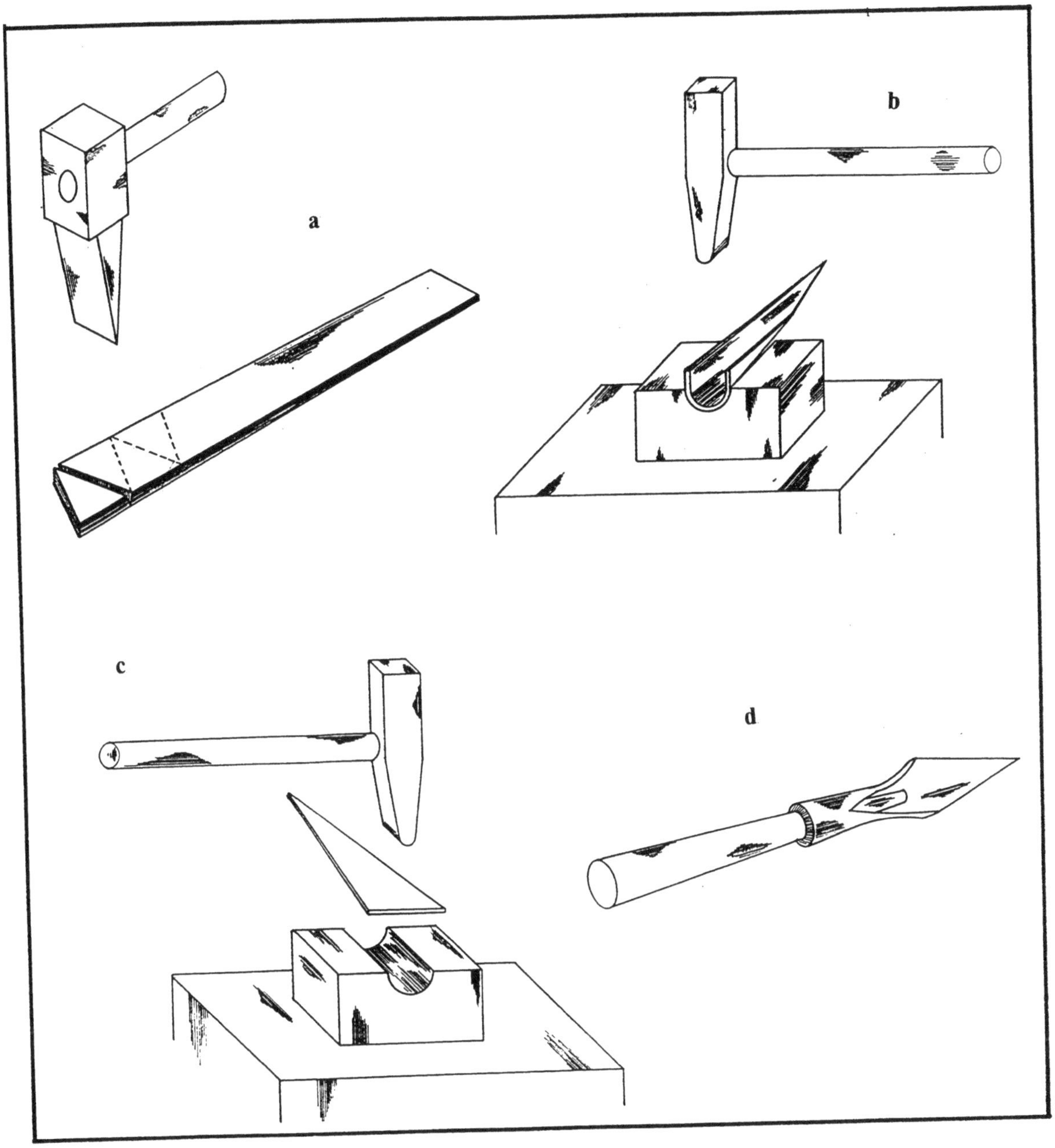

Figure 4.13 Stages of manufacture of a flat balista bolt head

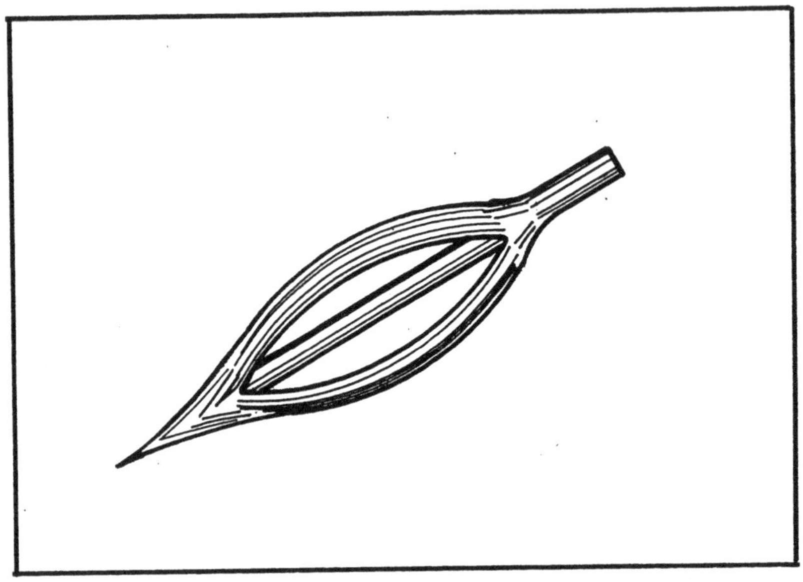

Figure 4.14 Fire arrow head

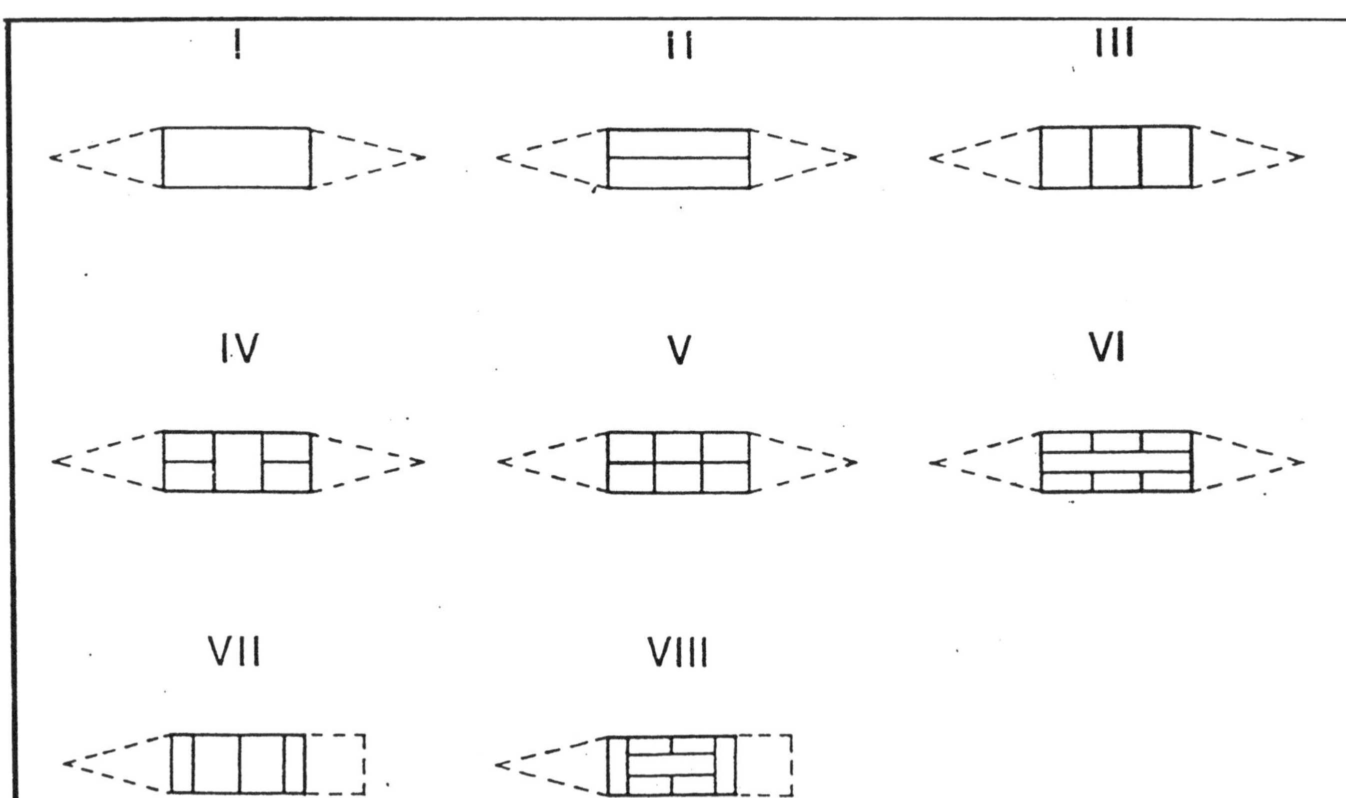

Figure 4.15 Various arrangements of central core and edging strips of a pattern-welded sword. (Gilmore *et al.* 1986)

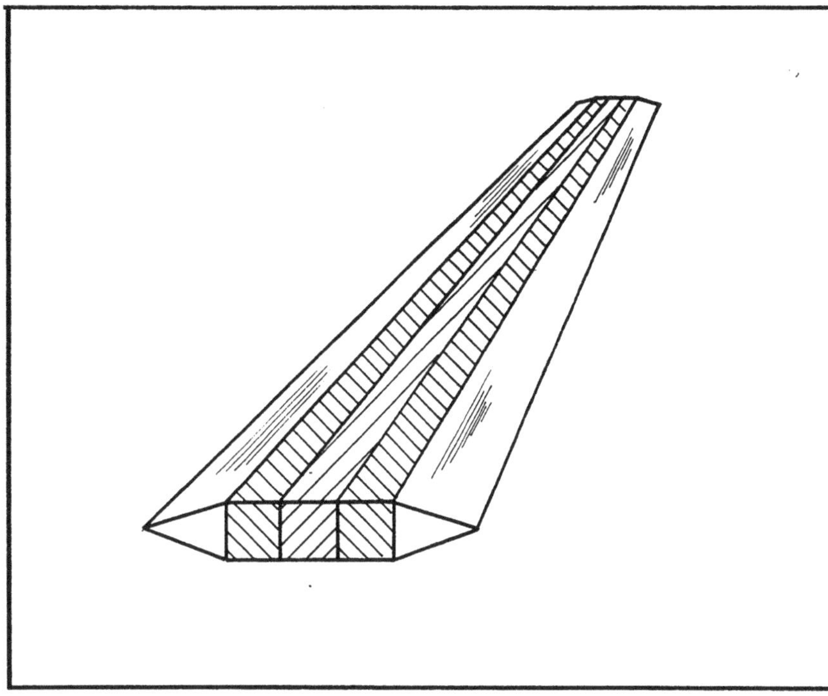

Figure 4.16 Central core and edging strips of a pattern-welded sword

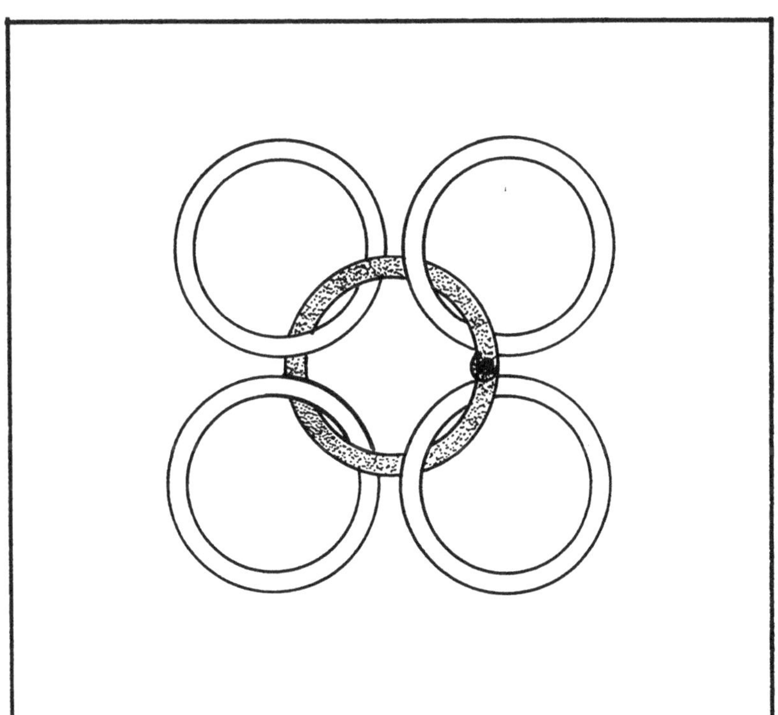

Figure 4.17 Solid rings joined by riveted ring

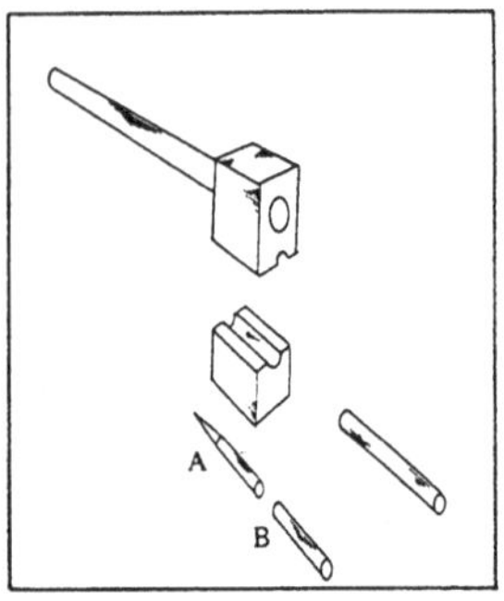

a Steel rod made between swages.

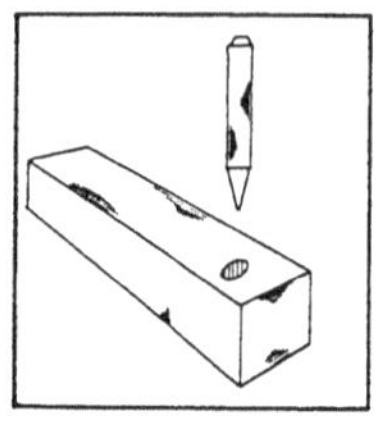

b Tapered punch used to punch a hole.

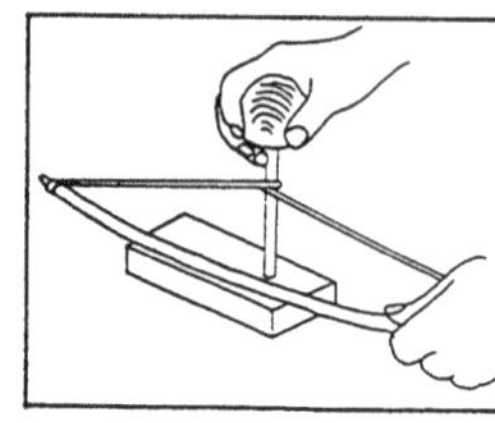

c Bow drill used for lapping punch into die.

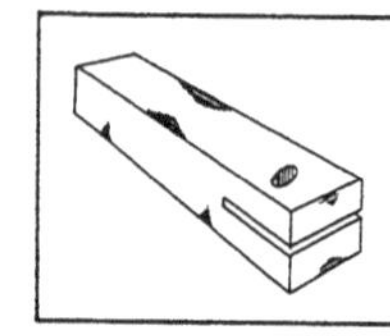

d Slot cut into the end of the bar.

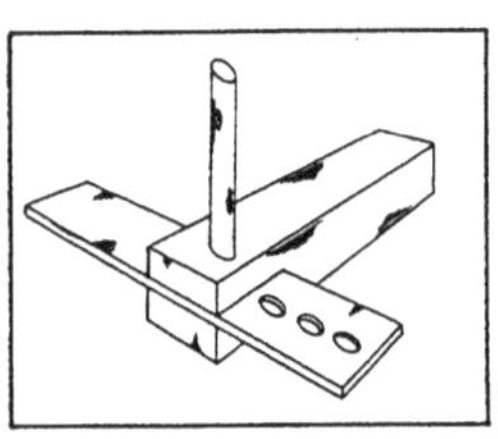

e Punch and die with the strip in place.

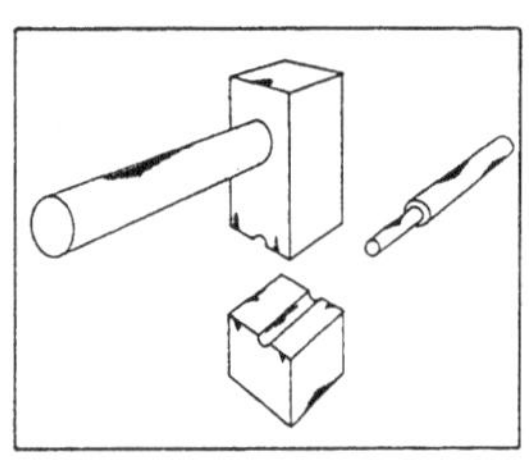

f Swage forming stepped punch.

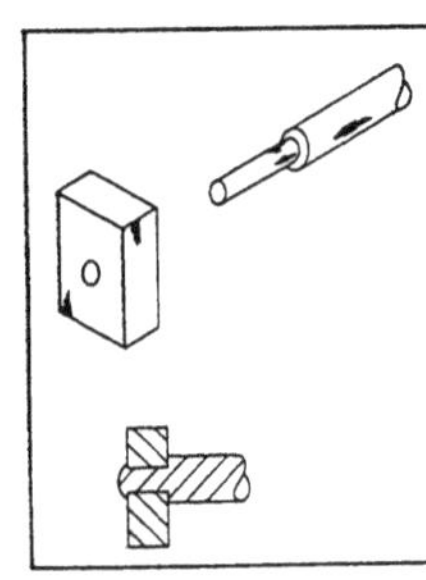

g Tenon.

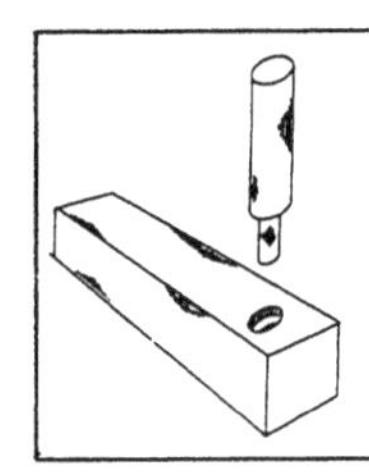

h Stepped punch used to make a stepped hole.

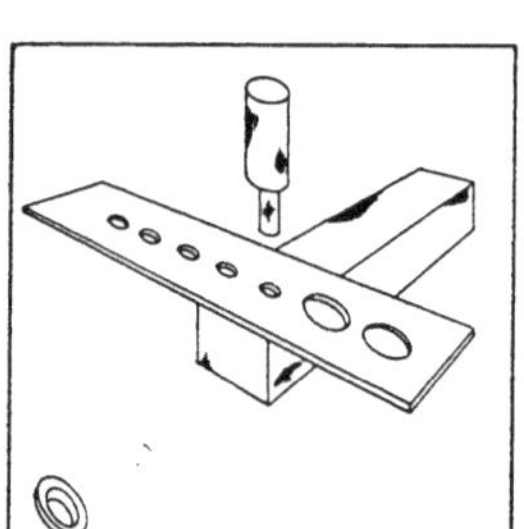

i Punch and die in use.

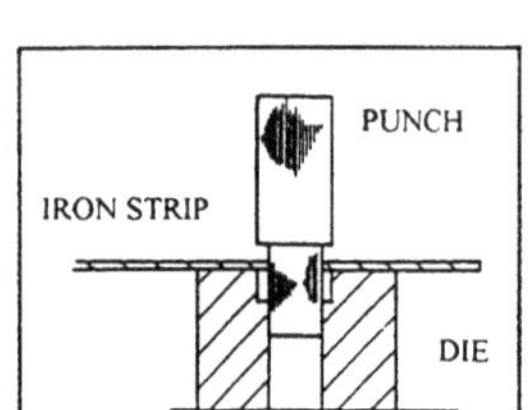

j Method of locating punch in die.

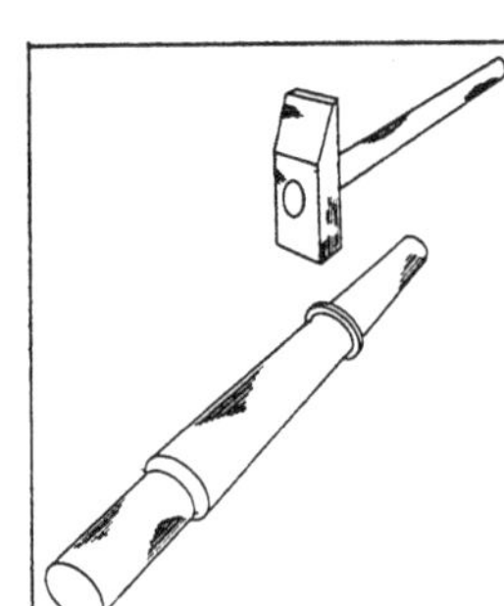

k Ring hammered on a tapering mandrill.

Figure. 4.18 Stages in producing a punch and die

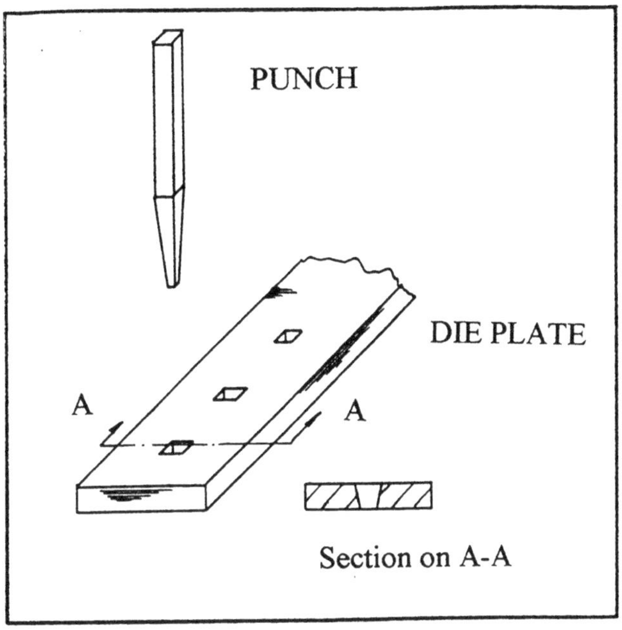

Figure 4.19 Experimental draw plate and punch

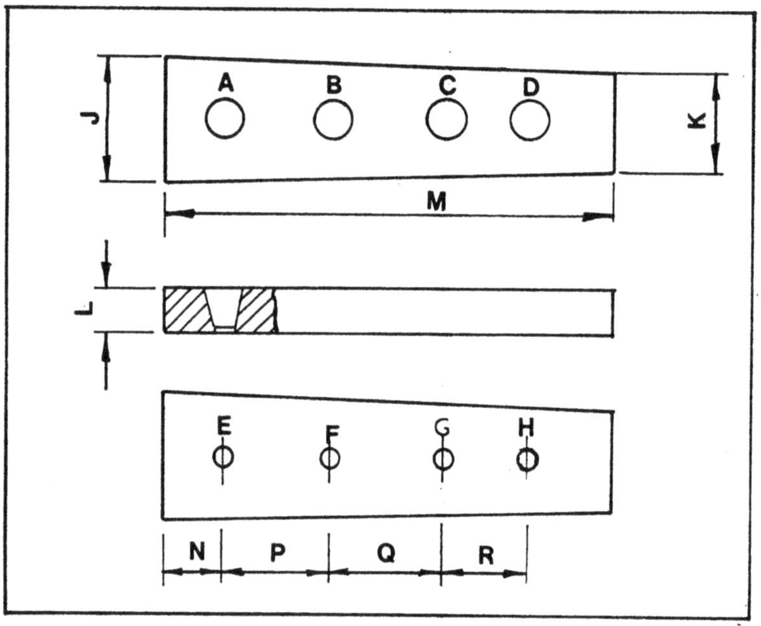

Figure 4.20 Schematic view of drawplate from Vindolanda

CHAPTER 5

Evaluation of Microscopic Debris from Bloomsmithing and Blacksmithing Experiments

And catch the burning sparks that fly
Like chaff from the threshing-floor.
Longfellow, The Village Blacksmith

5.1 Introduction

So far in this book we have considered two major forms of archaeological evidence which tell us about the way in which the Romans processed and manufactured items with iron: chapter 3 considered the aspects of bloomsmithing using a furnace, the design of which was based on the remains of those found in Britain; and, chapter 4 considered the various techniques which would have been used to manufacture a range of everyday items, from styli to swords.

A by-product of every processing and manufacturing stage which might all to easily be overlooked is the hammer scale. This chapter considers four main questions concerned with the production of hammer scale and other debris.

(i) Are there morphological and chemical differences in the hammer scale from the different activities carried out at a smithy i.e. bloomsmithing and blacksmithing; or indeed between blacksmithing processes?

(ii) How does the hammer scale produced in the experiments compare with that found in the archaeological record?

(iii) Is there a pattern to the way in which hammer scale is distributed about a forge depending upon the process involved?

(iv) How does the hammer scale produced in twentieth century forges compare with what was produced in the experiments and in the archaeological record?

Consideration of these questions will allow us to determine if examination of "waste" material in the archaeological record will indicate the use of the site. The experiments studied were those involving bloomsmithing (chapter 3) and artifact production (chapter 4). Archaeological hammer scale from the excavations of the Basilica at Silchester (Richards, forthcoming) was used as a comparison for the experimentally produced hammer scale. Five working forges were considered for comparison of modern hammer scale with that produced in the Roman era.

Numbering system

It should be noted at this point, that the numbering system for the experiments in this chapter have been altered from those used in chapters 3 and 4. This only affects new number 15 and 16. The new system is presented in table 5.1.

New Numbers	Description of Experiment	Comments
1	Bloom iron to workable iron	
2	Bloom iron to workable iron	
3	Bloom iron to workable iron	
4	Bloomsmithing	
5	Bloomwelding	
6	Drawing down	Forging strips
7	Drawing down	
8	Drawing down	No distribution map
9	Drawing down, with striker	No distribution map
10	Drawing down	No distribution map
11	Bloomwelding	
12	Bloomsmithing	No distribution map
13	Bloomwelding, in tongs	No distribution map
14	Bloomwelding, with striker	No distribution map
15	Bloomwelding, in tongs	Was EXP 11 in chap 3
16	Bloomwelding, in tongs	Was EXP 12 in chap 4

Table 5.1 Numbering system for chapter 5

5.2 The formation of hammer scale

The term hammer scale is applied to both macroscopic and microscopic residues associated with iron-making and iron-working. The hammer scale discussed in this chapter is the residue that has been produced by two separate and distinct processes, those of bloomsmithing and blacksmithing. McDonnell (in Ottaway 1992, 471), has suggested that artifact production is a two-stage process. The first is smelting, and the second blacksmithing. This is an oversimplification, because it implies that the refining of a bloom (bloomsmithing) is simply a blacksmithing process. It is true that superficially the two processes have a lot in common, in that they both involve heating and hammering. This is where the similarity ends. The temperatures and the materials are different and are worked in different ways. This suggests that they may produce different types of hammer scale. It seems a more realistic approach to consider the manufacture of artifacts as a three-stage process, those of smelting, bloomsmithing and blacksmithing. The process of bloomsmithing is simple and straightforward and should produce hammer scale which is consistent in chemistry and morphology. Blacksmithing is a more complicated process which may produce hammer scale that is unique to each process since processes are conducted at different temperatures and the temperature is one of the factor that governs the type of debris which forms.

Hammer scale from bloomsmithing

Hammer scale is an unavoidable consequence of heating and hammering iron. When bloomsmithing is carried out many factors will influence the form and chemistry of the hammer scale. By considering these stages of production individually it is hoped to create a clear picture of what takes place. The factors which must be considered are:

(a) The bloom in the fire
(b) The bloom in transit from the fire to the anvil
(c) The bloom resting on the anvil
(d) The area directly under the hammer
(e) The area under the hammer and resting on the anvil
(f) The area around the hammer but not under the face
(g) Shock-waves through the mass and their effect.

(a) The bloom in the fire

When the bloom is in the fire, it is in an oxidising atmosphere, which will cause the surface layers to become oxidised. Differential expansion between the iron and the iron oxide will cause some surface flakes to become detached from the body. These will fall into the bottom of the fire and possibly combine with other impurities from the bloom, such as slag. Slag adhering to the surface of the bloom will become liquid and run off the bloom into the bottom of the fire. Slag present near the surface of the bloom and which has a means of exit will also melt and run out of the bloom through the effects of gravity and capillary action. As the slag melts, larger pieces of iron that have been held in place by the slag will no longer be secure and will also become detached. These will fall into the bottom of the fire and re-oxidise.

(b) The bloom in transit from the fire to the anvil

The bloom will be held in tongs with enough force to prevent it from falling. No matter how lightly it is held, the gripping force will cause some material to become dislodged from the bloom while it is in transit. The outer layers will be in contact with the air and this will cause further oxidisation and the displacement of some flakes.

(c) Resting the bloom on the anvil

As the bloom comes to rest on the anvil the latter will cause chilling of the surface of the bloom where it has been in contact. Once again, differential expansion will dislodge some of the surface flakes.

(d) The area under the hammer

When the first hammer blow meets the surface of the iron, the surface layer under the hammer face will chill because of the temperature differential between the hammer surface and the iron. The chilling and consequent contraction will cause the oxide film to be dislodged from the body of the iron. The surface of the hammer will chill this scale so it will not weld into the iron. As the hammer rises the natural elasticity of the bloom will cause it to try to return to its original shape, and this action will push these pieces of scale off the surface. The compression in one area of the bloom will cause expansion in other areas, and this will cause the surface layers, which are colder, to break away from the hot metal underneath.

(e) The area under the hammer and resting on the anvil

During the first few heats of the bloom, the surface will be irregular, and at some places on the bloom there will only be point of contact. When the hammer descends, the area under the hammer and on the anvil will be forced flush with the surface of the anvil, causing chilling of that area. Shockwaves will be transmitted through the bloom and their vibrations will also cause the surface layers to become loose.

(f) The area around the hammer but not under the face

The area under the hammer is compressed and the surface layer is pulled downwards as the hammer head sinks into the iron. This stretching of the surface will cause fracturing of the surface layer at the interface of the oxide layer and the body of the iron.

(g) Shock-waves through the mass of iron and their effect

The shock-waves from the hammer blows will travel through the bloom and cause vibrations that will cause the cooled surface layers to dislodge.

Characteristics of Hammer Scale

The surfaces of hammer scale flakes can be either "moulded" or "free". The moulded surface may be the result of a hammer blow which will mould the surface under the hammer to the form of the face of the hammer and the other surface will be moulded to the shape of the surface it rests on. Free surfaces will form as a result of heating, or the effect of a force causing the flake to dislodge. Flakes will dislodge close to the hammer face but not underneath it. These may have a free surface on both sides, one from the effects of heating, the other from being separated from the parent body.

The above processes are common to both bloomsmithing and

blacksmithing but there are additional factors which are particular to each process.

Bloomsmithing

Bloom iron is a mixture of metallic iron and slag. Both the iron and the slag will contribute to the type of hammer scale produced when the bloom is smithed. The bloom starts with a high concentration of slag present, perhaps as much as 10% or more (Wynne and Tylecote 1958, 190). Hammering expels large quantities of slag, so any hammer scale produced by this process will comprise iron scale, scale which is exclusively slag, and scale which is iron with a high concentration of slag. With each hammer blow some slag will be completely expelled from the mass, but other particles of slag will be trapped at or near the surface. These will be dislodged during the next heating and will either fall off into the fire, or become dislodged after extraction from the fire. This dislodging is caused by a combination of differential expansion and the effects of shock waves. Alternatively, some scale may be expelled by the force of the hammer blow.

Blacksmithing

Blacksmithing is performed on iron from which most of the slag has been removed. The majority of the hammer scale from blacksmithing should be iron oxide with very small quantities of slag.

Formation of the shapes of hammer scale

An excellent description of the formation of hammer scale has been provided by Allen (1986, 97-8) and is repeated here to provide further information concerning some of the other factors which influence its formation.

> "The first few strokes of the hammer upon the raw bloom after its first reheat will therefore send an abundant spray of liquid slag droplets over the area around the anvil and hearth. Sufficiently small droplets, and those taking long enough atmospheric paths, will freeze before touching the ground and other surfaces, and so should assume a spheroidal form dictated chiefly by surface tension. More complex shapes might result where droplets had either acquired a spin, or proved unstable in flight because of their large mass. Droplets that fail to freeze in flight will, in varying degrees, either (a) mould to the shape of the surface on which they land, (b) alter in form due to impact, or (c) break up on landing to yield secondary splatter, which may then repeat either (a) or (b). The outcome may be a uniformly thin flake, a bun-like mass, or a dimpled spheroid, and will depend chiefly on the momentum and viscosity of the slag at impact. Meanwhile the bloom slowly cools as it is further worked. The slag now expelled from it will become increasingly viscous and rich in crystallised iron oxides. Irregular masses of crystal mush, rather than spheroidal drops, should provide the main microscopic residue. Such fragments may finally freeze either during flight, or upon touching a surface.
>
> By no means all the slag present in the bloom at its first reheat will be thus lost. Some may remain within the mass, and will only be expelled by hammering only after further reheats. Continued hammering as the bloom gradually cools to below the freezing point of the slag should dislodge thin films of slag in the shape of thin scales, one face of which will preserve the local form and, perhaps, the surface texture of the bloom. However, few of these scales will have had contact with the hammer.
>
> What of the fate of the liquid slag touched by the rapidly descending hammer? Some will chill against the much cooler hammer, and so be torn off or even violently extruded between hammer and bloom. Such rapidly cooled and sheared films should carry, on one or both surface, delicate striae created by asperities present on, particularly, the face and edges of the hammer".

The section above deals with the likely formation of slag. When smithing a bloom, the scale produced from pure iron has also to be considered. Much of the later stages of bloomsmithing are conducted at welding heat (see glossary). At this temperature the metal is at its most plastic stage and some of the surface, although not actually liquid, is fluid enough to act as a liquid if subjected to pressure. Carbon lowers the melting point of iron, and it is possible that some of the surface layers of iron will absorb carbon from the fuel, thus forming tiny droplets of carbon-rich iron. The melting point of these droplets may be low enough for them to become liquid. Other parts of the surface will be semi-liquid iron, and may act in the same way as slag, and will probably form into the same shapes. This hammer scale is likely to be of a different chemical composition to slag hammer scale. Slag hammer scale is comprised of non-iron particles from the iron ore which have become trapped during smelting.

There may be several different ways in which slag can be expelled from a bloom. The bloom is a constantly changing shape and hammering may bring slag close to the surface and also provide it with a channel of escape. If slag has been brought close to the surface and has a means of escape, it can melt while the bloom is in the fire, run out, and be picked up again when the bloom is reintroduced into the fire. It is uncertain if there is any difference in the finished shape of the slag.

5.3 Procedure for the study of hammer scale

The hammer scale investigated in the experiments was studied in terms of its morphology (shape), chemical composition and distribution. The area of the forge where the experiments were conducted was marked off into squares 0.2 m x 0.2 m, as shown schematically in figure 5.1. At the end of each experiment or series of experiments, scale was carefully collected from each of the marked squares, weighed and kept for further analysis. Collection of the hammer scale was made using a small stainless steel (non-magnetic) dustpan as shown in figure 5.2.

It was decided owing to the large number of samples to study in depth only representative squares from the whole grid area. The squares marked * in Figure 5.1, that is D5, F3, F5, I6 and J5 were arbitrarily chosen as representative of the whole. Squares D5, I6 and J5 tended to be moderately covered, whilst F3 and F5 collected greater quantities of scale, yet were far enough away from the anvil to not be filled with heavy lumps falling off the parent material.

Morphology

The morphology of the scale was performed using a binocular microscope at low magnification.

Each of the samples for the five squares (D5, F3, F5, I6 and J5) were taken and put through a series of sieves of graded mesh size, four in all: 5.60 mm. 2.00 mm, 0.60 mm and 0.212 mm. This provided a set of five samples (one for each mesh size and a final set smaller than 0.212 mm). The sieved samples were then examined under the microscope to determine the ratio of flakes, spheres and other material. This examination method has two advantages:

(i) it is quick and easy to use; and,

(ii) a similar method has previously been used by (Allen 1986) which will allow comparison of results.

The results for each sieve size were expressed as an ordinal (0-5) given by the area occupied by the different elements seen in the field of view of the microscope as follows:

5 >50% (dominant)
4 >30% but <50% (common)
3 >15% but <30% (rare)
2 >5% but <15% (very rare)
1 >1% but <5% (extremely rare)
0 (none present)

Chemical composition

The chemical composition of the scale was determined using XRF(X-ray fluorescence) analysis. Similar material has been analysed previously using this method (Fulford and Allen 1992), and sample preparation is quite straightforward. This analysis requires a sample size of at least 2.5 grammes and so it was not possible in all cases to make use of this technique. Where only a small quantity of material for analysis was available, a scanning electron microscope (SEM) was used. A draw-back of the SEM is that it examines only a very small part of the sample material, whilst the XRF method analysis the whole sample. The SEM does, however have the advantage of allowing a detailed inspection of the natural surface.

Distribution

Using the data obtained for the weights of samples collected from the different grid squares it was possible to develop "scale distribution maps" for the different experiments. This data was normalised (expressed as a percentage of the total) to allow comparison of the different processes.

5.4 Morphology of hammer scales

As has already been discussed in section 5.2, it is not unreasonable to expect that the different processes of bloomsmithing and blacksmithing will produce both morphologically and chemically different hammer scales. Further it is possible that there are differences between hammer scales from different blacksmithing processes. For example, the scale from a forge-welded operation may well be different from that produced when forging a billet down to a bar.

The manufacture of a billet is conducted at welding heat, around 1200°C. The straightforward forging of a hammer head needs no welding and can be conducted at a temperature of 900°C. The production of a pattern-welded sword is a combination of welding and forging. Welding requires temperatures of 1200°C, while forging is conducted at 900°C. It is possible that temperatures and working methods for different activities may affect the morphology and chemistry of hammer scale.

Six different experiments were chosen to see if there was any marked visual difference between the types of hammer scale produced. These were:

(a) Bloom iron forged into a billet by traditional methods. The blacksmith worked alone. (Experiment 3)

(b) Bloom iron welded inside a pair of tongs. The blacksmith worked alone. (Experiment 5)

(c) Wrought iron forged into a blank suitable for making a copy of a hammer from Silchester. The blacksmith was assisted by a striker. (Experiment 9)

(d) The wrought iron blank was forged into a hammer. The blacksmith worked alone. (Experiment 10)

(e) A block of wrought iron was forged into a strip. (no distribution map)

(f) A pattern-welded sword was made. The blacksmith worked alone. (Experiment 16)

All the samples were passed through the set of four sieves. In the following figures the scale of the photographs the magnification varies as stated. In general, the magnification shown may be achieved with a hand magnifying glass. Thus the photographs may be used in the field to help with the preliminary evaluation of hammer scale.

Definition of flake and spheres

For the purposes of this book, a flake will be considered as a piece of material whose thickness is small compared with its cross sectional area. The mathematical definition of a sphere is a shape where every point on the surface is equally distant from its centre. In the context of this book the "spheres" are spheroidal rather than true spheres.

Other slag material

In addition to hammer scale, there are two other types of material common to almost all of the experiments and the scale from Silchester - a black glassy substance, and nodules of a white substance. Both were examined and analysed with the SEM. The white substance was identified as predominantly calcium and silicon in varying amounts, and the black glassy substance was found to be iron and silicon in varying proportions. Full details of the analysis can be found in sections 5.5 and 5.6. The black glassy material which occurs as spheres and flakes is common to all experiments and the hammer scale collected from modern forges.

Discussion of photographs of hammer scale

The hammer scale collected has been photographed. For each of the experiments outlined above (a-f), a set of photographs shows hammer scale for the different sieve sizes. Not all sieve sizes are shown in each case, and in most cases, the samples from sieve sizes 0.212 mm and <0.212 mm are not considered. This is because at this size to the naked eye, or under low magnification they show negligible difference.

Scale of size 5.60 mm and larger will be considered in detail along with a general discussion of the samples from the other sieve sizes.

Bloom iron forged into a billet using traditional methods

Figure 5.3 shows the hammer scale produced when bloom iron was forged into a billet by traditional methods (Experiment 3). The scale is almost exclusively flat flakes. There are no spheres. At the smaller sieve size of 2.00 mm, a small number of spheres are present, but again the scale is almost exclusively flakes. One of the larger scales is shown in detail in Figure 5.3 a. It has a shiny bubbly surface texture, and a small sphere can be seen on the surface of the flake. The shiny "free" surface suggests that this flake came from the outside of the parent material. The bubbles are caused by gas which is forced out of the bloom by heating and hammering. (The gas forms spheres in much the same way as soap bubbles are formed. The larger spheres tend to be hollow and thus fragile, whilst smaller spheres are more robust.) The other side of this flake, (Figure 5.3 b) has a more "moulded" surface with a torn top edge, part of which shows depressions which are from the gas bubbles (and correlate with the spheres on the other side.). The hammer scale would have been forced from the parent bar before the spheres were fully formed. Cooling of the flake (either by removing from the heat or "freezing" during contact with hammer or anvil) would also prevent the full formation of the sphere The blister (in the right hand bottom corner) on the surface of the flake in Figure 5.3a shows the remains of a gas bubble. Figures 5.3 b and 5.3 c show both sides of a typical flake the shiny "free" outer surface (Fig 5.3 a) and "moulded" inner surface (Fig 5.3 b) are clearly visible.

Bloom iron welded inside a pair of tongs

Figure 5.4 shows the hammer scale resulting from bloom iron being welded inside a pair of tongs (Experiment 5). At size 5.6 mm and above, the scale is essentially flake like, but there are more nodules than produced in the bloom welding by the traditional, unconfined method, and some spheres. These features are also found in the scale produced at the smaller sieve size of 2 mm (Figure 5.4 a). It is interesting to note the different aspect ratio of these two type of flakes, (ratio of thickness to area) which is probably caused by the containment of the bloom and prolonged contact with a cold block used in tong bloom welding (see Figure 3.27).

Wrought iron forged into a blank for hammer production

Figure 5.5 shows hammer scale from wrought iron which was forged into a blank for making a copy of a hammer from Silchester. In this instance, the parent material was a block of well-purified wrought iron. The flakes are very large and of a uniform thickness. The outer layer is smooth and represents the outer skin of the block, which would have oxidised evenly over the uniform surface to produce large flakes. One sphere is visible (see the bottom right hand corner of Figure 5.5 a) its presence is unexpected, and is possibly due to contamination from the forge. At smaller sieve sizes the scale is exclusively flakes with no spheres (Figure 5.5 b).

Wrought iron blank forged into a hammer

The wrought iron blank discussed in section above, was then forged into a hammer. The scale produced during this process is shown in Figure 5.6 and 5.6 a. As might be expected, the scale flakes are similar to those shown in Figures 5.5 and 5.5 a. The scale in the largest sieve size (compare Figures 5.5 and 5.5 a) is smaller, because the "parent bar" by this stage was also smaller. The working temperatures of the two operations were similar (in the range 850°C - 1000 °C).

Block of wrought iron was forged into a strip

A block for wrought iron approximately 50 mm square in section was forged down into thin strips about 2 mm thick and 12 mm wide. The strip forged in this process was subsequently used in the piles for the pattern welded sword. Figures 5.6 and 5.6 a show the hammer scale collected at the two largest sieve sizes.

This scale is very similar to that produced in the previous two experiments (see Figures 5.5, 5.5 a, 5.6 and 5.6 a), wrought iron forged into a blank for a hammer and the forging of the hammer itself. The temperature in this process would have been slightly higher that the other forging processes (approx. 1000 °C). (A higher temperature will allow the material to be more readily reshaped, although at the expense of increased oxidation of material – more hammer scale.) The larger flakes would have been produced in the early part of the forging process, when the block was quite large and the smaller scale later on. In the centre of Figure 5.6 a large sphere is visible, and one or two others a just showing. It is probable that the higher working temperature has allowed their formation.

Manufacture of a pattern-welded sword

Figure 5.7 shows the hammer scale from the welding of bundles of strips into a sword blank. This operation would have been carried out at welding heat (1200 °C). During this process large quantities of material are lost due to excessive oxidisation, the residue produced has a low aspect ratio, and may more properly be described as nodular. One large piece of scale which was dislodged from the sword blank towards the end of the forge welding operation is shown in Figures 5.7 a and 5.7 b. Figure 5.8 a is a view of the "free surface" where the division between the bundles of different bundles of iron can just bee seen. Figure 5.8 b shows the inside of the flake, with the division of the bundles is clearly visible.

Figures 5.7 c and 5.7 d show hammer scale collected at the end of the forging of the sword blank. At the larger sieve size, Figure 5.7 c, there are flakes with no spheres present, whilst at a sieve size of 2.0 mm, a number of spheres may be seen (Figure 5.7 d). The small spheres would have been produce during forge welding.

Conclusion

At a size of 5.6 mm or larger any differences in hammer scale should be apparent to the naked eye. This will help in the preliminary analysis of a forge site. The hammer scale from forging a billet is made up of flakes with usually one "free" and one "moulded" surface. The moulded surface may have what appear as bubbles on the surface, although the formation of free spheres tends not to occur. The hammer scale produced by welding inside a pair of tongs is in the form of nodules, with a small proportion of flakes which have one free and one moulded surface.

When forging without welding is conducted, the hammer scale takes the shape of large flakes where the moulded surface is very smooth and has few indications of bubbles. When a bar of wrought iron is forged with no welding, similar flakes are produced which are smaller in area – which is to be expected as the parent bar is smaller. When a sword was made (this involved considerable welding) much hammer scale was produced which took the form of nodules rather than flakes.

At 2.00 mm the differences are less noticeable to the naked eye, but generally follow the same forms as was discussed for sieve size 5.6 mm. At sieve sizes less than 2.0 mm, there are scarcely any differences visible to the naked eye and a more powerful magnification is necessary for differences to be seen. However, microscopic examination of the smaller sieve size material shows it to be of the same appearance as the larger scale for each process studied here. The occurrence of spheres in any sample is an indication that the material may have been worked at a high temperature. This is difficult to quantify but it is probably in the range of a welding heat (1200°C) which is well above the heat required for general forging (850°C - 1000°C).

5.5 XRF analysis of hammer scale

Analysis is required to establish if there is a chemical difference between the hammer scales produced from different types of forging operations. Three experiments were selected to cover the three processes most commonly carried out in a forge:

(a) forging refined iron with no welding;

(b) forging a raw bloom by traditional methods; and,

(c) forge welding.

Pieces of hammer scale were collected from the smithing debris using a magnet and a sample comprising several of these pieces was prepared for XRF analysis on a Philips PW 148 X-Ray Spectrometer running Philips X 40 software. A wide range of international standards was used for calibration, including GSP1, DRN, FER 1, FER 2, BCS 301, BSC 393, 603-1, 606-1 and 612-1.

Interpretation of the data

Each of three samples were analysed for each sieve size, and found to be composed of essentially two compounds: iron oxide (Fe_2O_3) and silicon oxide (SiO_2). Iron oxide is essentially hammer scale, whilst the silicon oxide is produced by the slag.

Table 5.2 presents the percentage content of the two compounds in the debris for the different sieve sizes. This data is presented graphically in Figure 5.8. In all cases the ratio of slag : hammer scale is generally constant for the range of sieve size considered. However, for the less pure bloom at the smallest sieve size the debris tends towards a slag to scale ratio of 50 : 50. Experiment (a) which involves the processing of a well refined billet (with little slag remaining) produces mostly iron oxide and little silicon oxide residues. Samples from experiments (b) and (c) which involve less well refined iron contain more silicon based slag and hence contain more the silicon oxide.

Sieve size (mm)	Experiment (a)		Experiment (b)		Experiment (c)	
	SiO_2	Fe_2O_3	SiO_2	Fe_2O_3	SiO_2	Fe_2O_3
5.60	2.54	95.91	23.11	72.41	30.48	64.63
2.00	3.45	94.01	21.50	74.61	30.55	64.47
0.60	2.84	95.09	23.61	71.85	32.36	62.48
0.212	2.74	95.66	29.25	66.00	42.81	50.71
<0.212	3.35	94.51	45.34	49.52	44.03	48.14

Table 5.2 % content of SiO_3 and Fe_2O_3 in produced debris

5.6 SEM analysis of debris

The XRF analysis of hammer scale has shown it to be predominantly FeO_3 (see chapter 5.5). There are two other commonly occurring materials a black glassy substance and white nodules.

Both materials had been observed in the hammer scale from experimental work as well as in the sample of hammer scale from Silchester. It seems quite likely that these residues are a feature occurring with hammer scale, and as such may be relevant to the identification and isolation of hammer scales from different processes.

Specimens were mounted on aluminium studs and then were either gold or carbon coated. They were examined in a JELL model JXA 840 Scanning Electron Microscope and micro-analyses. (Micro-analysis was carried out using a model AN 10000 link analyses using energy dispersive-analysis.)

Note: samples have to be coated with either gold or carbon before they can be examined, therefore, where gold (Au) is shown on the spectrum it should be discounted. Samples that show no Au were coated with carbon, which does not show up on the spectrum.

Black glassy material

To look at, the black glassy material specimens are not identical in appearance, a difference which is reflected in their different chemical composition. The XRF analyses of the black glassy samples show that they are similar, in that they are composed of large amounts of iron (Fe) with silicon (Si) in varying degrees.

Figure 5.10 shows the analysis of the material taken from the Basilica at Silchester. It may be seen that the sample is made up of iron and silicon with calcium (Ca) and potassium (K) as the other main constituents. Figures 5.11 - 5.15, 5.16 and 5.16 a show micro analyses and photographs of other samples of black glassy material produced in the experimental work. Figure 5.15 which shows analysis of material taken from an experiment involving the welding of a billet has a similar chemical composition graph to the sample from Silchester. Another sample of the glassy black material, a sphere - which can be seen in Figure 5.16, also has a similar composition to those shown from Silchester and billet welding samples, although in appearance it is very different (compare Figures 5.10 and 5.15 with 5.16 and 5.16 a). The Silchester and billet welding samples have an irregular porous appearance, whilst the sphere has a much smoother surface. The fine crack which can clearly be seen in Figure 5.16 a indicates that this material is brittle.

Comparison of the chemical analysis from these three samples with the other black glassy material (from experiments in traditional bloom forging (Figure 5.11); bloom welding in tongs (Figure 5.12); and, forging iron (Figure 5.13)) seems on first inspection to suggest that they are completely different. The main difference is the abscence of calcium (Ca) and potassium (K) and a much reduced silicon (Si) content. It will be seen in the next section that silicon and calcium are the main constituents of the white lumps, and so it is not unreasonable therefore to suggest that there is some "contamination" by this material present in the "black glassy" samples, which would cause the difference in chemical analysis. Indeed, the bloom welding sample (Figure 5.15) was taken from a piece of black slag with white flakes. Hence the chemical difference between samples may be as much in the quality of the iron being forged as in the manner of forging.

It is interesting to note that the two bloom welding experiments, by traditional methods (Figure 5.11) and in tongs (Figure 5.12) produce glassy material which is very similar in appearance and chemical composition. Figure 5.14 show the analysis of a second piece of black material from the same forging experiment as that shown in figure 5.13, the two traces are very similar, as could be expected.

White lumps

The white material analysed in the SEM was found to be either silicon (Si), calcium (Ca) or a combination of the two in varying proportions. The sample from the Basilica at Silchester is almost exclusively silicon (Figure 5.17), whilst the samples from the experimental work are mostly calcium (Ca) with some silicon. Figures 5.18 – 5.21 show the white lumps of material and their corresponding SEM analysis for the debris collected during the experimental work. The analysis of the white lumps from the traditional bloom forging experiment (Figure 5.18) showed that the material consisted of calcium and silicon in a ratio of about 3:1, whilst results from the experiment welding iron into a billet shows a calcium : silicon ratio of 1:1 (Figure 5.19). Finally the sample from the experiment forging iron is predominantly calcium (Figure 5.20).

The origin of the white material is not certain, but it is likely to be either from the material being forged or the forge (lining) itself. It is unlikely that the material will have come from the fuel as this is charcoal which is essentially carbon (as are wood and coal). Silicon is a main constituent in sand stone, as is calcium in lime stone. If lime were used as a flux, this too would introduce calcium to the forging process.

5.7 Hammer scale in reproduction experiments

This section describes and discusses the hammer scale distribution in terms of the morphological variation found over the chosen representative grid squares (see section 5.3) and in terms of the total mass distribution around the blacksmith's anvil. In the broadest possible sense, hammer scale can be divided into two main types: flakes and spheres, with a large number of sub-divisions in each type (Allen 1986) for the purposes of this analysis the scale will be divided into flakes and spheres, with the additional classes of "black slag" and "white lumps" (see section 5.6).

As already stated, the tables which follow the hammer scale from each of the five representative grid squares on the smithy floor has been passed through the series of sieves, and for each size the ordinal abundance for each of the four main sub-divisions of types of hammer scale noted.

Experiments which involved the same process have been taken together: experiments 7, 8 and 9 (Table 5.3) involved blacksmithing with a striker, 1, 2 and 4 (Table 5.4) involved bloomsmithing by traditional methods and 12,13 and 16 (Table 5.5) were bloom welding inside tongs.

The tables have been arranged such that each column groups the five representative grid squares for each type of experiment, whilst reading across the tables allows comparison of the same grid square for each process. It can be seen that in the three processes; bloom welding (in tongs); bloomsmithing (traditional method); and, blacksmithing with a striker, flakes dominate all the sieve sizes. At the two largest sizes, 5.60 mm and 2.00 mm, the occurrence of any other material is extremely rare. Most of the spheres, black slag and white lumps occur at 0.600 mm and below. In experiments 7, 8 and 9, which were blacksmithing, there are few spheres and overall the occurrence of white lumps is extremely rare - in fact only one lump occurs. This experiment was conducted at a temperature of approximately 900°C. There should be very little slag in the wrought iron, and that will be inside the structure, and so not in a position to be liberated by heat.

The black slag is made up mostly of iron and silicon (in oxide form, see section 5.6). This means that it will probably be a brittle material which will shatter on impact either from the hammer or on contact with the floor. This may well account for it only being found at the smaller sizes. The white lumps are mostly calcium and silicon (in oxide form, see section 5.6) which are brittle materials and again will shatter when subject to force – as before – this may account for it only being found at smaller sizes. The spheres are probably formed when the temperature of the iron is near welding heat, where the material has slag which can be liberated from the mass of iron. Although spheres occur in all experiments, they are more prevalent in the bloom welding experiments where the forging temperature is near to welding heat (1100°C - 1200°C).

Grid Ref. I6

Sieve Size (mm)	Flakes	Spheres	Black Slag	White Lumps
5.60	5	0	0	0
2.00	5	0	1	0
0.600	4	2	1	3
0.212	4	2	1	3
<0.212	4	2	1	1

Grid Ref. J5

Sieve Size (mm)	Flakes	Spheres	Black Slag	White Lumps
5.60	5	0	1	0
2.00	5	0	1	0
0.600	5	1	1	1
0.212	5	2	1	1
<0.212	4	1	1	1

Grid Ref. F3

Sieve Size (mm)	Flakes	Spheres	Black Slag	White Lumps
5.60	5	0	0	0
2.00	5	0	1	0
0.600	5	2	1	0
0.212	5	2	1	0
<0.212	4	2	1	1

Grid Ref. D5

Sieve Size (mm)	Flakes	Spheres	Black Slag	White Lumps
5.60	5	0	0	0
2.00	5	2	1	0
0.600	5	2	1	1
0.212	5	2	1	1
<0.212	5	1	1	1

Grid Ref. F5

Sieve Size (mm)	Flakes	Spheres	Black Slag	White Lumps
5.60	5	0	0	0
2.00	5	0	1	0
0.600	5	2	1	1
0.212	5	2	1	1
<0.212	5	3	1	1

Table 5.3 Experiments 7, 8 and 9. - Blacksmithing with a striker

Grid Ref. I6

Sieve Size (mm)	Flakes	Spheres	Black Slag	White Lumps
5.60	5	0	2	0
2.00	5	1	1	0
0.600	5	2	2	1
0.212	4	3	2	2
<0.212	4	1	1	1

Grid Ref. J5

Sieve Size (mm)	Flakes	Spheres	Black Slag	White Lumps
5.60	5	0	0	0
2.00	5	1	1	0
0.600	5	2	1	1
0.212	4	3	1	1
<0.212	3	1	0	0

Grid Ref. F3

Sieve Size (mm)	Flakes	Spheres	Black Slag	White Lumps
5.60	5	0	1	0
2.00	5	1	1	0
0.600	5	2	1	1
0.212	4	2	1	1
<0.212	4	1	0	1

Grid Ref. D5

Sieve Size (mm)	Flakes	Spheres	Black Slag	White Lumps
5.60	5	0	1	0
2.00	5	0	1	0
0.600	5	1	1	1
0.212	4	1	2	1
<0.212	4	1	1	1

Grid Ref. F5

Sieve Size (mm)	Flakes	Spheres	Black Slag	White Lumps
5.60	5	0	1	0
2.00	5	1	1	1
0.600	5	1	2	1
0.212	5	2	2	1
<0.212	5	1	0	0

Table 5.4 Experiments 1, 2 and 4 - Bloomsmithing traditional method

Grid Ref. I6

Sieve size (mm)	Flakes	Spheres	Black slag	White lumps
5.60	5	0	0	0
2.00	5	0	1	0
0.600	5	1	1	1
0.212	5	2	2	0
<0.212	5	1	1	0

Grid Ref. J6

Sieve Size (mm)	Flakes	Spheres	Black Slag	White Lumps
5.60	5	0	1	0
2.00	5	0	2	1
0.600	5	1	2	1
0.212	5	2	2	0
<0.212	5	1	1	0

Grid Ref. F3

Sieve size (mm)	Flakes	Spheres	Black Slag	White Lumps
5.60	5	0	0	0
2.00	5	1	1	0
0.600	5	2	1	1
0.212	5	3	1	1
<0.212	5	1	1	1

Grid Ref. D5

Sieve Size (mm)	Flakes	Spheres	Black Slag	White Lumps
5.60	5	0	0	0
2.00	5	1	1	0
0.600	5	1	1	1
0.212	5	1	1	1
<0.212	5	1	0	1

Grid Ref. F5

Sieve Size (mm)	Flakes	Spheres	Black Slag	White Lumps
5.60	5	0	0	0
2.00	5	0	1	0
0.600	5	2	1	0
0.212	5	1	2	1
<0.212	5	1	1	1

Table 5.5 Experiments 12, 13 & 16 welding bloom iron in tongs

Hammer scale distribution maps for experiments

Hammer scale throughout the smithy was assessed by weighing the scale collected in each of the grid squares (see figure 5.1). Normalised hammer scale distribution maps are presented for the following experiments 1-7, 10, 11, 13, 15 and 16, which cover a range of processes. For all of these distribution maps the blacksmith was right handed and moved between G3 and H3 facing G8. The smith worked both with and without a striker.

Interpretation of hammer scale distribution maps.

Experiment 1 – forging bloom iron into workable iron

Figure 5.21 shows the distribution of hammer scale as a percentage of the total scale collected. The area shown in the thick line is the area where the largest deposit of hammer scale fell. It can be seen that the largest concentration is to the top left hand side of the anvil in grid reference H 5. The blacksmith was right handed and was working alone. Hammer blows arrive at an angle to the surface being struck, from an observation of the way the blacksmith worked the hammer scale seemed to come off at an angle of 45° to the left of the anvil which is where the largest concentration of hammer scale falls. The hammer scale forms a roughly circular pattern about the anvil. Large particles will heavier and will not travel far from the anvil, smaller particles will travel further. It is to be expected that there will be a greater concentration of material closer to the anvil.

Experiment 2 - forging raw bloom

This was welding small pieces of bloom iron inside tongs. The reason for using tongs was that the iron was contained and when struck with a hammer it will only be able to compress in on itself.

Figure 5.22 shows the distribution of hammer as a percentage of the total scale collected. The largest concentration is to the top left hand side of the anvil in grid reference G 5.

Hammer blows arrive at an angle to the surface being struck. The scale will be forced off the parent bar to the left side of a right-handed blacksmith and to the right side of a left-handed blacksmith. From observation of the way in which the right-handed blacksmith worked, the hammer scale was seen to come off at an angle of 45° to the left of the anvil and the largest concentration of hammer scale was here although the centre of concentration was further away from the anvil. This is probably because the blacksmith is right-handed. This is explained by the fact that in this experiment considerably more force was used to deliver the hammer blows. In Experiment 1, the hammer scale only extended from E to I and from 1 to 7 on the grid. In Experiment 2, the hammer scale is found from B to J and from 2 to 8, once again forming a roughly circular pattern.

The distribution of particles will depend on the initial velocity imparted to them by the hammer blow. The shape will also play a role in the distance individual particles will travel. This is a matter of aerodynamics and has not been considered here. It is to be expected that there will be a larger concentration of material closer to the anvil.

Experiment 3 – forging bloom iron into workable iron

Figure 5.23 shows the distribution of hammer scale as a percentage of the total scale collected. The area shown in the thick line is the area where the largest deposit of hammer scale fell. The amount of force used in experiment 3 was similar to that used in experiment 2 and the range of hammer scale distribution is approximately the same. Again it forms a roughly circular pattern with the centre being G 5-6.

Experiment 4 – forging bloom iron into workable iron

Figure 5.24 shows the distribution of hammer scale as a percentage of the total scale collected The area shown in the thick line is the area where the largest deposit of hammer scale fell.

In this experiment bloom iron was welded in tongs. The amount of force used in experiment 4 was similar to that used in experiment 3 and the range of hammer scale distribution is approximately the same. Again it forms a roughly circular pattern with the centre being a little lower at G 5.

Experiment 5 – welding bloom iron inside tongs

Figure 5.25 shows the distribution of hammer scale as a percentage of the total scale. The area shown in the thick line is the area where the largest deposit of hammer scale fell.

In this experiment bloom iron was welded in tongs. The amount of force used in experiment 5 was similar to that used in experiments 3 and 4 and the range of hammer scale distribution is approximately the same. Again it forms a roughly circular pattern with the centre being a little lower at G 4-5.

Experiment 6 - drawing billet to bar

Figure 5.26 shows the distribution of hammer as a percentage of the total scale collected. In this experiment steeledwrought iron was forged into a bar by a blacksmith and one striker. The amount of force used was considerably larger than in any previous experiment but the range of hammer scale distribution is approximately the same. Again it forms a roughly circular pattern with the centre being at G 5-6. This is not what was expected. When a blacksmith works alone the largest concentration of scale is at grid reference G5, so it was thought that a blacksmith working with a striker would produce a larger concentration at H4. The concentration near the anvil is due to the striker brushing the hammer scale off the anvil towards himself.

Experiment 7 – drawing billet to bar

Figure 5.27 shows the distribution of hammer scale as a percentage of the total scale. The area shown in the thick line is the area where the largest deposit of hammer scale fell.
This is a repeat of experiment 6 the range of hammer scale

distribution is approximately the same. Again it forms a roughly circular pattern with the centre being at G 5-6.

Experiment 10 - forging wrought iron into bar

Figure 5.28 shows the distribution of hammer scale as a percentage of the total scale. The area shown in the thick line is the area where the largest deposit of hammer scale fell. This is a repeat of experiments 6, 7, 8 and 9. The results are similar but there are some differences. The range of hammer scale distribution is approximately the same. Again it forms a roughly circular pattern but the range is smaller in the major axis of the ellipse and centre has moved to G 6. This is probably due to the striker changing his position.

Experiment 11 - welding pieces into a billet

Figure 5.29 shows the distribution of hammer scale as a percentage of the total scale collected. The area shown in the thick line is the area where the largest deposit of hammer scale fell. This experiment investigated bloom welding in tongs. The results are similar to most of the previous experiments, but there are some differences. The range of hammer scale distribution is approximately the same. Again it forms a roughly circular pattern but the range is smaller in the major axis of the ellipse and the centre has moved to G 6. This is probably due to the striker changing his position.

Experiment 13 - bloom welding in tongs

Figure 5.30 shows the distribution of hammer scale as a percentage of the total scale. The area shown in the thick line is the area where the largest deposit of hammer scale fell. This is welding small pieces into a billet. The results are different to most of the previous experiments. The range of hammer scale distribution is approximately a circle with the centre at G 4.

Experiment 15 – bloom cut with axe then welded inside tongs

Figure 5.31 shows the distribution of hammer scale as a percentage of the total scale. The area shown in the thick line is the area where the largest deposit of hammer scale fell. This experiment also investigated bloom welding in tongs. The results are similar to most of the previous experiments, but there are some differences. The range of hammer scale distribution is approximately the same. Again it forms a roughly circular pattern but the range is smaller in the major axis of the ellipse, and the centre has moved to G 4-5.

The general pattern of distribution is an ellipse. This is to be expected because the force of the hammer blow is not directed vertically down but at an angle which will tend to exert more force sideways. The body of the smith will stop scale coming towards him from reaching the end of its flight, but his feet will disperse it sideways and forwards.

The greatest concentration of scale is to be found in column G between rows 4 and 6 predominantly in row 5. It was expected that a blacksmith and striker working together would produce a different pattern of distribution, but the factor of the striker cleaning the anvil and sweeping the scale towards himself meant the scale was still concentrated in roughly the same area. It is probable that the striker would brush the scale to a part of the floor away from his own feet which could be to either side of the anvil. There were two different strikers who assisted; both were right-handed and used their right hand to brush the scale off when the iron was reheating in the fire. They had not been given any specific instructions concerning the scale on the anvil and performed the cleaning in the same manner of their own accord. The maximum spread can be seen in Experiment 6, where the hammer scale is distributed between columns C to K. In this experiment the striker used much more force than was necessary to perform the task in hand and later experiments show that as he became used to the work much less force was used. In the majority of experiments the scale is distributed between Column D to J.

Experiment 16 – making a pattern-welded sword

Figure 5.32 shows the distribution of hammer scale as a percentage of the total scale collected. The area shown in the thick line is the area where the largest deposit of hammer scale fell. This is bloom welding in tongs. The results are similar to most of the previous experiments but there are some differences. The range of hammer scale distribution is approximately the same. Again it forms a roughly circular pattern but the range is smaller in the major axis of the ellipse and centre has moved to G 5.

The effect of "handed-ness" on the distribution maps

Cross cultural studies put the incidence of right-handed-ness at around 90% i.e. 90 people out of every 100 (Springer and Deutsch 1982, 142), and a variety of indirect evidence suggests that this has been the case as far back as prehistoric times (Dennis 1958; Dart 1949; Uhrbrock 1973; and, Coren and Porac 1977). In the first instance it is reasonable to assume that the majority of forges that have (and will be) excavated were manned by right handed blacksmiths. Other things being equal, the pattern of hammer scale distribution recorded in the experiment will be similar to those for smithies in the archaeological record. In the instances where the blacksmith was left handed it would be expected that the hammer scale distribution map would be a mirror image of that for the right handed smith. Further varieties/complications will exist if the smith is working with a striker who may or may not be left handed. If we assume that eligibility for a striker's job is completely unaffected by handedness (i.e. a right handed blacksmith will not want only e.g. a right handed striker) then the likelihood of a blacksmith and striker both being right handed is 81% i.e. 81 instances out of 100. Again, this is high enough for it not to be unreasonable to assume that in the first instance both the blacksmith and striker were right handed.

5.8 Hammer scale in modern forges

A survey of hammer scale was undertaken for five modern forges, (one of which was the author's forge in which all the reproduction experiments had taken place). The survey of modern forges will give information on:

(i) whether the hammer scale produced in modern forges has similarities with that produced in the experiments on antiquity;

(ii) the variation in morphology and distribution of hammer scale between forges;

(iii) the similarity of hammer scale between forges

The forges chosen for the study represent a range of locations in Britain. An additional consideration in selection was the problem of the influence of modern welding equipment. This can produce material (specifically spheres) very similar in appearance to those produced by fire welding. Hence smithies were chosen where there was no "modern equipment" or where it was kept away from the forging site.

Many factors contribute to the debris which will be found on a forge floor. The factors that can be readily observed include: the type of metal being forged; the type of fuel used; the composition and surface character of the floor; and, the type of work being carried out in the forge. Other factors such as the smith's methods of forge management and his personal working habits are less easy to quantify and may give rise to differences in both morphology and distribution of scale in the forge.

The following forges were surveyed with the permission of their owners.

Name of forge: Plantation Forge
Name of blacksmith: Mr John Anstee
Location: 56, Plantation Avenue, Arnside, Westmorland, LA5 0HX
Type of building: Wattle and daub
Type of floor: Beaten earth
Metals used: Wrought iron, Brass
Fuel: Charcoal
Type of work: Ancient artifact reproduction

Name of forge: Unnamed
Name of blacksmith: Mr Les Eaton
Location: Blists Hill Open Air Museum, Madeley, Ironbridge Gorge Museum. Shropshire.
Type of building: Brick
Type of floor: Concrete
Metals used: Wrought iron
Fuel: Charcoal
Type of work: Decorative iron work

Name of forge: Ironstone Forge
Name of blacksmith: David Sim
Location: 1 Melrose Avenue, Reading, Berks, RG6 7BN
Type of building: Single story, prefabricated concrete, with pitched roof
Type of floor: Concrete
Metals used: Mild steel, wrought iron, brass, copper, lead
Fuel: Coke
Type of work: Decorative iron work, general iron work, tool-making

Name of forge: Unnamed
Name of blacksmith: Mr Ivan Smith
Location: Woodside, Sneads Green, Nr Droitwich, Worcestershire, WR9 0PY
Type of building: Brick
Type of floor: Beaten earth
Metals used: Mild steel and alloy steels
Fuel: Coke
Type of work: Decorative iron work, general iron work, tool-making

Name of forge: Tibbett, Son and White
Name of blacksmith: Mr George Tibbett
Location: Merchants Place, Reading, Berkshire
Type of building: Wood frame with brick infill
Type of floor: Concrete
Metals used: Mild steel, brass, copper, lead
Fuel: Coke
Type of work: Decorative iron work, general iron work, tool-making

5.9 Analysis of floor debris from modern forges

The analysis of the samples was performed in the same way as for the experimental forging, as described in section 5.3. Since all the smithies examined were working forges however, access was restricted and it was not possible to mark the whole floor out in the grid used for the experimental work.

Based on experience from the earlier work it was decided to collect data from a row of grid squares D5-J5, (refer Figure 5.1) as this would give a representative range of samples, whilst minimising disruption in the smithy. By selecting the area on the side of the anvil opposite the smith, the falling scale would not be disturbed or damaged by pedestrian traffic.

In the case of the two smithies which had earth floors, some of the samples contained so much soil that it was impossible to get any clear indication of what they contained at the two smallest sizes of sieve. At these sizes, hammer scale is hardly visible to the naked eye and in all probability would be overlooked by archaeologists during an excavation. It is possible to separate hammer scale using a magnet but it was necessary to collect non-magnetic scale from these samples as well as magnetic scale. Owing to these difficulties it was decided to leave the two smaller grid sizes out of the analysis. It was not deemed necessary to quantify the material that was not hammer scale because the working practices of each smith would affect the amounts of other waste produced.

In the tables which follow (tables 5.6-5.10) the samples for the seven grid squares investigated have been analysed for the three largest sieve sizes. Also presented is a brief description of the other material found in the samples

The elements found in the hammer scale are abbreviated as follows.

F = flakes S= spheres BS = black slag WL = white lumps

Grid reference D5

Sieve size (mm)	F	S	B S	W L	Other material
5.60	0	0	1	0	Stones
2.00	5	0	1	0	Stones & charcoal
0.60	5	1	1	1	Stones & charcoal

Grid reference E5

Sieve size (mm)	F	S	B S	W L	Other material
5.60	5	0	0	0	Stones & charcoal
2.00	5	1	1	1	Stones, charcoal & soil
0.60	5	1	1	1	Stones, charcoal & soil

Grid reference F5

Sieve size (mm)	F	S	B S	W L	Other material
5.60	0	1	1	0	Stones & charcoal
2.00	5	1	2	1	Stones & charcoal
0.60	5	1	1	1	Stones, charcoal & soil

Grid reference G5

Sieve size (mm)	F	S	B S	W L	Other material
5.60	5	0	1	0	Stones
2.00	5	1	1	1	Stones & charcoal
0.60	5	2	2	1	Stones, charcoal & coal

Grid reference H5

Sieve size (mm)	F	S	B S	W L	Other material
5.60	5	0	0	0	Stones & coal
2.00	5	1	1	1	Stones, charcoal & coal
0.60	5	1	2	2	Stones & coal

Grid reference I5

Sieve size (mm)	F	S	B S	W L	Other material
5.60	5	0	0	1	Stones, charcoal & coal
2.00	5	1	1	1	Stones
0.60	5	1	2	1	Stones & coal

Grid reference J5

Sieve size (mm)	F	S	B S	W L	Other material
5.60	5	0	0	0	Stones, charcoal & coal
2.00	5	1	2	1	Charcoal & coal
0.60	5	1	1	1	Stones, charcoal & coal

Table 5.6 Analysis of debris from the forge of John Anstee

Grid reference D5

Sieve size (mm)	F	S	B S	W L	Other material
5.60	5	0	2	0	Wood & charcoal
2.00	5	0	3	0	Charcoal
0.60	5	3	3	1	Charcoal

Grid reference E5

Sieve size (mm)	F	S	B S	W L	Other material
5.60	5	0	2	0	Wood & charcoal
2.00	5	0	1	1	Iron off cuts
0.60	5	2	2	1	Charcoal

Grid reference F5

Sieve size (mm)	F	S	B S	W L	Other material
5.60	4	0	2	0	Wood & charcoal
2.00	5	2	3	0	Charcoal
0.60	5	1	1	1	

Grid reference G5

Sieve size (mm)	F	S	B S	W L	Other material
5.60	5	0	1	0	Charcoal
2.00	5	1	1	1	
0.60	5	1	1	0	

Grid reference H5

Sieve size (mm)	F	S	B S	W L	Other material
5.60	5	0	0	0	Wood & charcoal
2.00	5	1	1	1	
0.60	5	1	2	2	

Grid reference I5

Sieve size (mm)	F	S	B S	W L	Other material
5.60	5	0	3	0	Charcoal
2.00	5	0	3	0	
0.60	5	2	2	1	

Grid reference J5

Sieve size (mm)	F	S	B S	W L	Other material
5.60	5	0	2	0	Charcoal
2.00	5	1	1	1	
0.60	5	2	3	1	

Table 5.7 Analysis of debris from the forge of Les Eaton

Grid reference D5

Sieve size (mm)	F	S	B S	W L	Other material
5.60	5	0	3	0	Iron off cuts, wood, charcoal & coke
2.00	5	1	3	1	Iron off cuts, wood, charcoal & coke
0.60	5	3	2	1	Wood & charcoal

Grid reference E5

Sieve size (mm)	F	S	B S	W L	Other material
5.60	5	0	2	0	Iron off cuts, wood, charcoal & coke
2.00	5	1	1	1	Wood, charcoal & coke
0.60	5	2	2	1	Wood, charcoal, coke, stone & dead insect

Grid reference F5

Sieve size (mm)	F	S	B S	W L	Other material
5.60	5	0	1	0	Iron off cuts, wood, charcoal & coke
2.00	5	1	2	1	Wood & charcoal
0.60	5	1	2	1	Wood, charcoal & stone

Grid reference G5

Sieve size (mm)	F	S	B S	W L	Other material
5.60	5	0	1	0	Iron off cuts, wood, charcoal & coke
2.00	5	1	2	1	Iron off cuts, wood & charcoal
0.60	5	2	1	1	Wood, charcoal & coke

Grid reference H5

Sieve size (mm)	F	S	B S	W L	Other material
5.60	2	0	5	3	Iron off cuts, wood, charcoal & coke
2.00	5	1	2	1	Wood, charcoal & lead
0.60	4	2	2	2	Wood & charcoal

Grid reference I5

Sieve size (mm)	F	S	B S	W L	Other material
5.60	5	0	0	2	Iron off cuts, wood, charcoal & coke
2.00	5	1	2	1	Wood, charcoal, coke & brass
0.60	5	1	2	1	Wood, charcoal, coke & brass

Grid reference J5

Sieve size (mm)	F	S	B S	W L	Other material
5.60	5	0	1	0	Iron off cuts, wood, charcoal & coke
2.00	5	3	2	1	Wood, charcoal, coke & brass
0.60	5	2	2	1	Wood, charcoal, coke & brass

Table 5.8 Analysis of debris from the forge of David Sim

Grid reference D5

Sieve size (mm)	F	S	B S	W L	Other material
5.60	5	0	2	1	Wood, charcoal, coke & leaves
2.00	5	2	1	1	Wood, charcoal, coke, leaves, twigs & stone
0.60	5	2	1	1	Charcoal & stone

Grid reference E5

Sieve size (mm)	F	S	B S	W L	Other material
5.60	5	1	2	1	Iron off cuts, wood, leaves & sand on large flakes
2.00	5	1	1	1	Wood & coke
0.60	5	2	2	1	Wood, charcoal, coke, leaves & stone

Grid reference F5

Sieve size (mm)	F	S	B S	W L	Other material
5.60	5	0	1	1	Iron off cuts, wood, charcoal, coke, leaves, stone & sand on large flakes
2.00	5	2	3	1	Wood, charcoal, coke & stone
0.60	5	1	1	1	Charcoal, coke & stone

Grid reference G5

Sieve size (mm)	F	S	B S	W L	Other material
5.60	5	0	1	0	Iron off cuts, wood, charcoal, coke & leaves
2.00	5	1	3	1	Borax, wood & charcoal
0.60	5	2	2	1	Charcoal, coke & stone

Grid reference H5

Sieve size (mm)	F	S	B S	W L	Other material
5.60	5	0	0	0	Iron off cuts, wood, charcoal & coke
2.00	5	1	1	1	Iron off cuts & wood
0.60	5	2	2	2	Charcoal, coke & stone

Grid reference I5

Sieve size (mm)	F	S	B S	W L	Other material
5.60	5	0	0	2	Iron off cuts, wood & coke
2.00	5	1	1	1	Wood, charcoal & coke
0.60	5	2	2	1	Charcoal, coke & stone

Grid reference J5

Sieve size (mm)	F	S	B S	W L	Other material
5.60	5	0	0	0	Iron off cuts, wood, charcoal & coke
2.00	5	3	2	1	Wood, charcoal & coke
0.60	5	2	2	1	Charcoal, coke & stone

Table 5.9 Analysis of debris from the forge of Ivan Smith

Grid reference D5

Sieve size (mm)	F	S	B S	W L	Other material
5.60	5	0	0	0	Iron off cuts, wood & coke
2.00	5	1	1	1	Wood, charcoal, coke & solder
0.60	5	1	1	1	Wood, charcoal, coke & solder

Grid reference E5

Sieve size (mm)	F	S	B S	W L	Other material
5.60	5	0	0	0	Iron off cuts, wood & coke
2.00	5	1	1	1	Wood, charcoal, coke & solder
0.60	5	2	2	1	Wood, charcoal, coke & solder

Grid reference F5

Sieve size (mm)	F	S	B S	W L	Other material
5.60	5	0	1	0	Iron off cuts, wood, coke & nail head
2.00	5	1	2	1	Wood, charcoal, coke, solder & dead insect
0.60	5	1	1	1	Wood, charcoal, coke & solder

Grid reference G5

Sieve size (mm)	F	S	B S	W L	Other material
5.60	5	0	1	0	Iron off cuts, wood & coke
2.00	5	1	3	1	Wood, charcoal, coke & solder
0.60	5	2	2	1	Wood, charcoal, coke & solder

Grid reference H5

Sieve size (mm)	F	S	B S	W L	Other material
5.60	5	0	0	0	Iron off cuts, wood & solder
2.00	5	1	1	1	Iron off cuts, wood, coke & solder
0.60	5	1	2	2	Wood & copper oxide

Grid reference I5

Sieve size (mm)	F	S	B S	W L	Other material
5.60	5	0	0	2	Iron off cuts, wood & coke
2.00	5	1	1	1	Wood, coke & solder
0.60	5	1	2	1	Wood, charcoal & coke

Grid reference J5

Sieve size (mm)	F	S	B S	W L	Other material
5.60	5	0	0	0	Iron off cuts, wood, charcoal, coke & solder
2.00	5	3	2	1	Iron off cuts, wood, charcoal, coke & solder
0.60	5	2	2	1	Wood, charcoal & coke

Table 5.10 Analysis of debris from the forge of George Tebbits

5.10 Discussion of analysis of floor debris

Flakes, spheres, black slag and white lumps

For all the forges investigated, the four basic components of hammer scale: flakes, spheres, black slag and white lumps were all present (tables 5.6-5.10). The most important factor is that the same material shows up in both modern forges and the experimental results. For each of the sieve sizes investigated in the modern forge samples with only one exception (H5 in table 5.8), the flakes predominate. This was also the case in the experimental studies (tables 5.3-5.5).

The same four components are present in the sample of hammer scale found at Silchester. Other materials found in the modern forges such as wood, stones, leaves and other metals were not found in the experimental samples. This was to be expected as great care was taken to exclude any contaminants to these samples.

The types of hammer scale produced in the experiments are the same as those produced in modern forges, and in some forges from antiquity.

As the sieve size decreases the quantity of spheres in each sample increases. Generally speaking, black slag is rare in the largest sieve sizes, but is the more common for the other smaller sizes.

White lumps are very rare at the large size but appear more often at smaller sizes.

Other material

The debris samples from the modern forges which were examined contained materials other than hammer scale including: wood, charcoal, lead, solder, brass, copper oxide, stone, leaves and soil.

In specialist workshops, it is expected that only a very few, specific materials would be found. The smithy of Mr. John Anstee is used solely for the forging of iron, and a study of the debris shows that the only metal present is iron, the other materials include charcoal and coal – the residue of the forge fuel. In a general blacksmith's workshop, repairs will form a significant part of the work, and a smith may well be asked to repair metal items which are not exclusively iron. A weather vane, for example, will have iron (or steel) for the main body, copper (for the vane), brass or bronze for the bearings, and gold in leaf form to gild the vane and compass points. Parts of all these materials could end up on the floor of the smithy. As a result, the floor of the smithy may well contain many different materials. Because of this sort of individual preference it is not possible to establish a general rule for extraneous material found in conjunction with hammer scale. Nevertheless its presence is significant as it will give indications of the type of work performed in the individual smithies.

Wood

Wood was found in all the samples taken but not at every sieve size. Wood is likely to be found in most forges and might come from a number of sources. Most blacksmiths will use wood for lighting the fire, irrespective of the fuel used for smithing. Wood is hammered on the anvil surface to break it into fine splinters which are then put in the bottom of the fire. Wood is cut into longer pieces on the anvil using a set in the hardie hole (see glossary). These are put on top of the smaller splinters.

A significant part of the blacksmith's work is putting wooden hafts onto tools for his own use, or on items made for customers. Most tools brought to the blacksmith for repair will require a new handle. Large shavings are produced from the initial removal of stock from the haft (these may well be used in lighting the fire). Small shavings will be produced from the finishing cuts to bring the haft to the correct size to fit the head of the tool.

At times, a wooden block is placed on the anvil and the hot iron laid on it and struck with a heavy wooden mallet (to straighten the bar without leaving hammer marks). This will produce splinters from the block and the mallet head as well as charcoal from the burnt block.

Charcoal

When charcoal is used as a fuel for forging, small amounts will be picked up from the fire when the iron is extracted, and will then be deposited on the floor. Some of the wood used as a fuel to light the fire will become charcoal in the fire. Blacksmiths using coal or coke as a fuel insert a large block of wood in the fire if they wish to keep the fire alive when they are not forging. Wood left like this becomes charcoal. Small pieces will remain in the fire, and will be picked up on the iron being forged.

Lead

Lead is used to make heavy hammers which are used when iron has to be hammered in a way that will not leave hammer marks on the surface, for example, when a twist needs to be straightened. A steel head will leave hammer marks and ruin the twist, but a lead hammer used in conjunction with a lead block will straighten the bar without leaving any marks. Lead blocks are also used on an anvil when decoration is applied to the surface of a piece of work. Small flakes of lead will become detached from the surfaces of the block when it is struck with a hammer, and many of these will fall to the floor. When hot iron is in contact with a lead block, some of the lead will melt and it may either drop onto the anvil or the floor, or may stick to the surface of the iron. When the iron contaminated with lead is struck with a hammer the lead may well be ejected from the surface of the iron and form residues with the same forms as hammer scale. This formation has been described by Allen (1986).

Solder

Although this is the material used by the tinsmith, most blacksmiths will use solder from time to time as a jointing medium, or to fill in small gaps or small holes. Solder was found in only one of the smithies examined, that of Mr. Tibbet. All the recorded samples were taken from around the anvil where Mr. Tibbet chose to conduct his soldering while the other smiths have special areas set aside for this purpose.

Brass

This alloy was found on only one site where the area in which the brass was worked is next to the anvil. In the other forges, the area for working non-ferrous metal was away from the anvil.

Copper

Only one piece of copper oxide was found.

Stone

Small pieces of stone form part of the floor of these smithies with beaten earth floors. Their abundance and similarity suggest that they were not imported onto the site on the blacksmith's shoes.

Leaves

The leaves have blown into the smithy from outside and are there by accident. (Discussion with the blacksmith revealed that leaves formed no part of his forging practices.)

5.11 Conclusion

The production of iron artifacts can be seen as a three-stage process: smelting, bloomsmithing and blacksmithing. Although prima facia the same in that they involve heating and hammering of a bloom, it was suggested that bloomsmithing and blacksmithing might produce different distinct residue (different raw materials and processing techniques) which could be used to interpret evidence from the archaeological record, as to the activities carried out at a particular site.

The XRF anaylsis has shown that there are chemical differences between the products of forging raw bloom, forging a refined bloom (artifact production) and welding. In general the purer the material being worked or processed, the purer the hammer scale produced. Conversely, the amount of silicon present in the hammer scale decreases in the purer material (Table 5.2). (In the experiments conducted, for forging the amount of Fe_20_3 is higher by a factor of 1.4 over smithing a bloom and by a factor of 1.6 over fire welding. The amount of $Si0_2$ present in forging is lower by a factor of 9.6 than that found in bloomsmithing and by 12.1 in fire welding.)

Together with the hammer scale, two other materials were present in most of the forging debris from both Silchester and the experimental work. These materials were: a black glassy substance which was found to be mostly iron oxide with varying smaller amounts of silicon (and calcium and potassium) depending upon the purity of the parent bloom; and, a white material which was found to be silicon and calcium in varying quantities. It seems likely that the black material comes from the slag in the iron bloom, whilst the white lumps have come from the furnace lining. It is reasonable to expect to find all of these types of forging debris in both the archaeological record and modern day forges.

The distribution maps of the hammer scale in the experiments have shown that the largest amount of formed in any experiment falls within the area with grid reference G5 and H5 (see Figure 5.1). The general distribution is in the form of a circle, centred just forward and to the left of a right handed blacksmith. Since right-handedness is found in 90% of the population, and has done since prehistoric times, it is not unreasonable to assume in the first instance that the distribution in other smithies will follow this distribution, The use of a striker does not seem to have a predominant effect on the overall distribution of the scale. Therefore it is suggested that the distribution of hammer scale in the experimental work will be similar to that found in most smithies.

There are some differences in the chemical composition, quantity and morphology of the hammer scale flakes and spheres between the different processes, particularly in fire welding (conducted at a high temperature) which produces a relatively large number of small spheres in the hammer scale.

In general, however, variation in the hammer scale products and distribution is a complex combination of the raw and furnace materials as well as the different processing techniques.

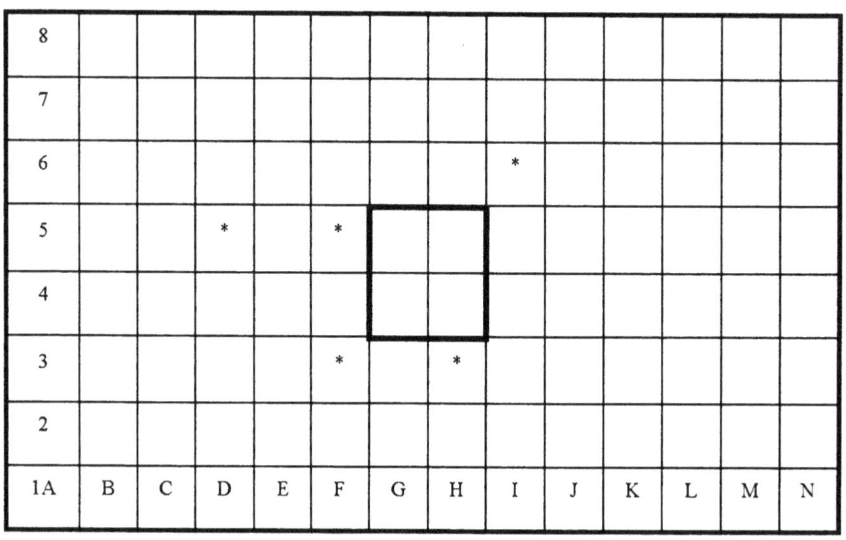

Figure 5.1 Floor plan of smithy (scale: grid size 20 cm x 20 cm.) The blacksmith stood at G3. The centre of the anvil is at G-H 4-5

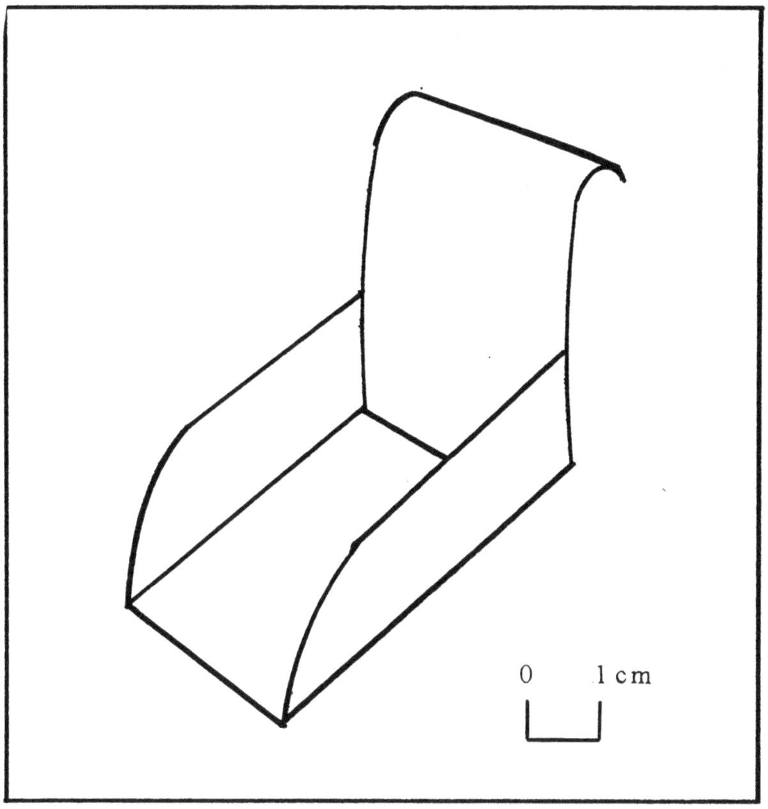

Figure 5.2 Pan for collecting hammer scale

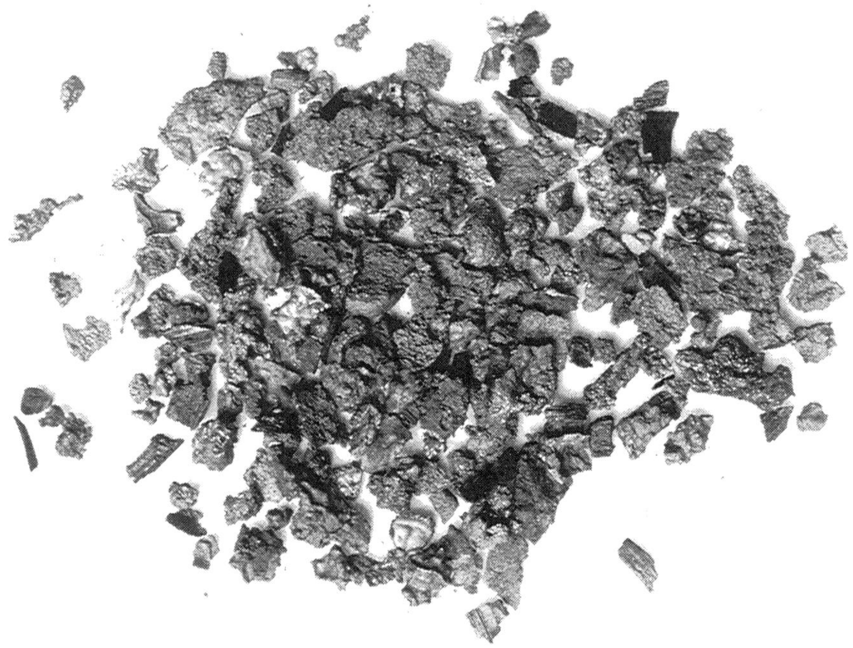

Figure 5.3 Hammer scale from bloom iron forged into a billet by traditional methods (magnification x 4). Sieve size 5.60 mm

Figure 5.3 a Hammer scale from bloom iron forged into a billet by traditional methods (magnification x 1.3). Sieve size 2.00 mm

Figure 5.3 b Detail of hammer scale from bloom iron forged into a billet by traditional methods (magnification x 4).

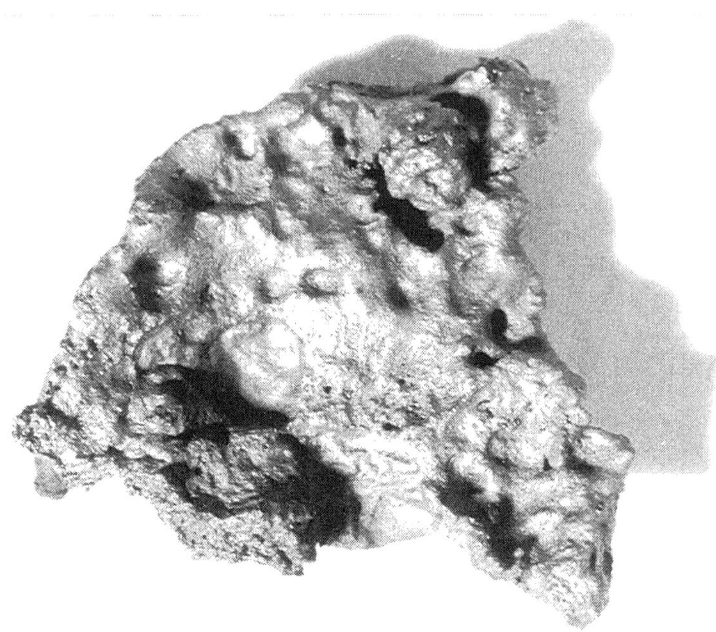

Figure 5.3 c Detail of hammer scale from bloom iron forged into a billet by traditional methods (magnification x 4).

Figure 5.4 Hammer scale from bloom iron being welded inside a pair of tongs (magnification x 1.3). Sieve size 5.6 mm

Figure 5.4 a Hammer scale from bloom iron being welded inside a pair of tongs (magnification x 1.3). Sieve size 2.00 mm

Figure 5.5 Hammer scale from wrought iron forged into a blank suitable for making a copy of a hammer from Silchester. (Magnification x 1.3) Sieve size 5.60 mm.

Figure 5.5 a Hammer scale from wrought iron forged into a blank suitable for making a copy of a hammer from Silchester. (Magnification x 1.3) Sieve size 2.00 mm.

Figure 5.5 b Hammer scale from wrought iron forged into a blank suitable for making a copy of a hammer from Silchester. (Magnification x 1). Sieve size 5.60 mm.

Figure 5.5 c Hammer scale from wrought iron forged into a blank suitable for making a copy of a hammer from Silchester. (Magnification x 1.3) Sieve size 2.00mm.

Figure 5.6 Hammer scale from forging wrought iron into strip (magnification x 1.3).Sieve size 5.6

Figure 5.6 a Hammer scale from forging wrought iron into strip (magnification x 1.3).Sieve size 2. 0

Figure 5.7 Hammer scale from welding of bundles of strips into a pattern welded sword blank (Magnification x 2)

Figure 5.7 a Free surface of large hammer scale from welding of bundles of strips into a pattern welded sword blank (Magnification x 4)

Figure 5.7 b Moulded surface of large hammer scale from welding of bundles of strips into a pattern welded sword blank

Figure 5.7 c Hammer scale from welding of bundles of strips into a pattern welded sword blank (Magnification x 1) Sieve size 5.60 mm

Figure 5.7 d Hammer scale from welding of bundles of strips into a pattern welded sword blank (Magnification x 1) Sieve size 2.00 mm

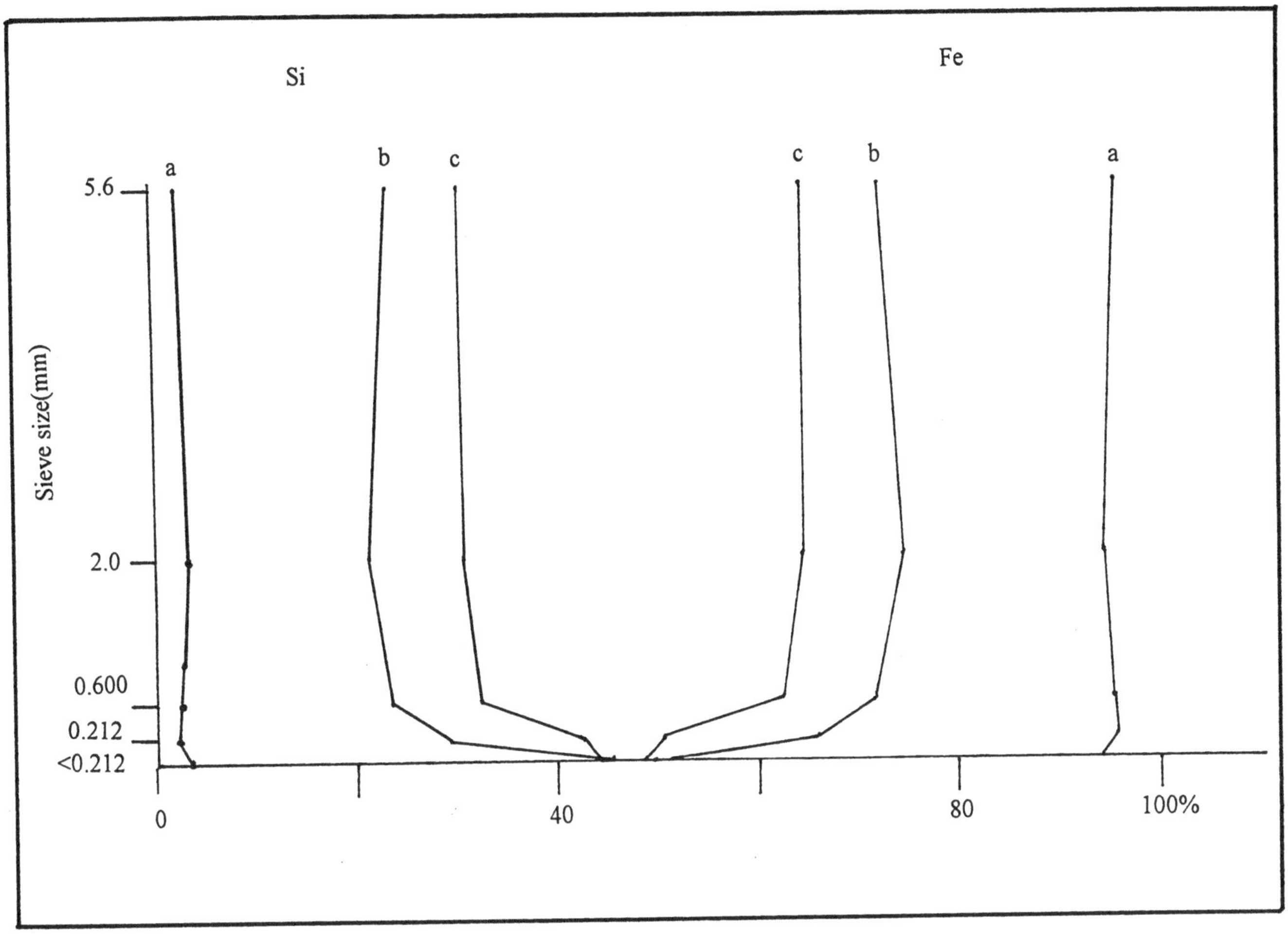

Figure 5.8 Graph shows Fe_2O_3 against SiO_2

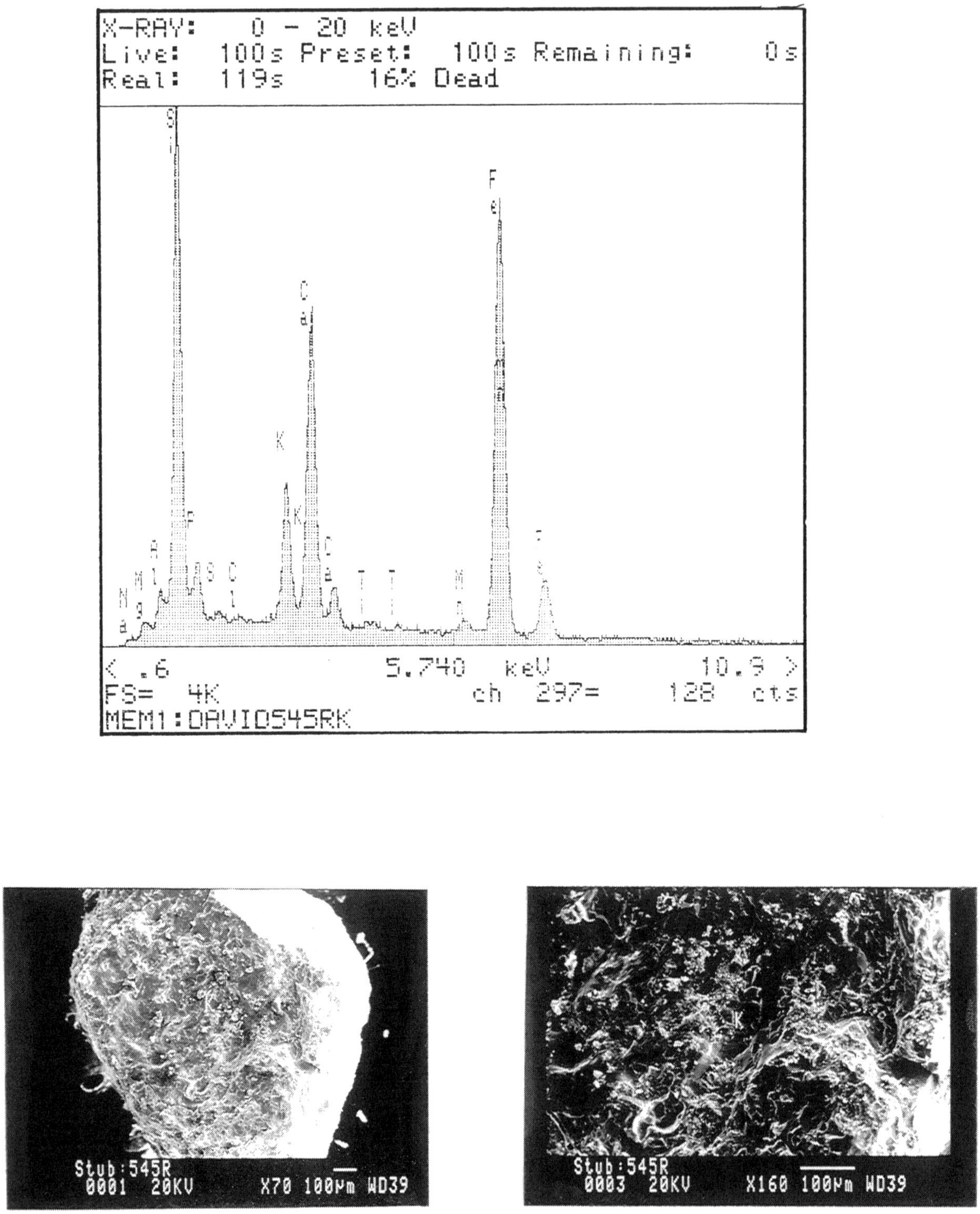

Figure 5.9 Black glassy material from the Basilica at Silchester.

Figure 5.9 shows the sample from the Basilica at Silchester. The analysis shows it to be made up of iron and silicon in almost equal proportions, with calcium and potassium the other major constituents.

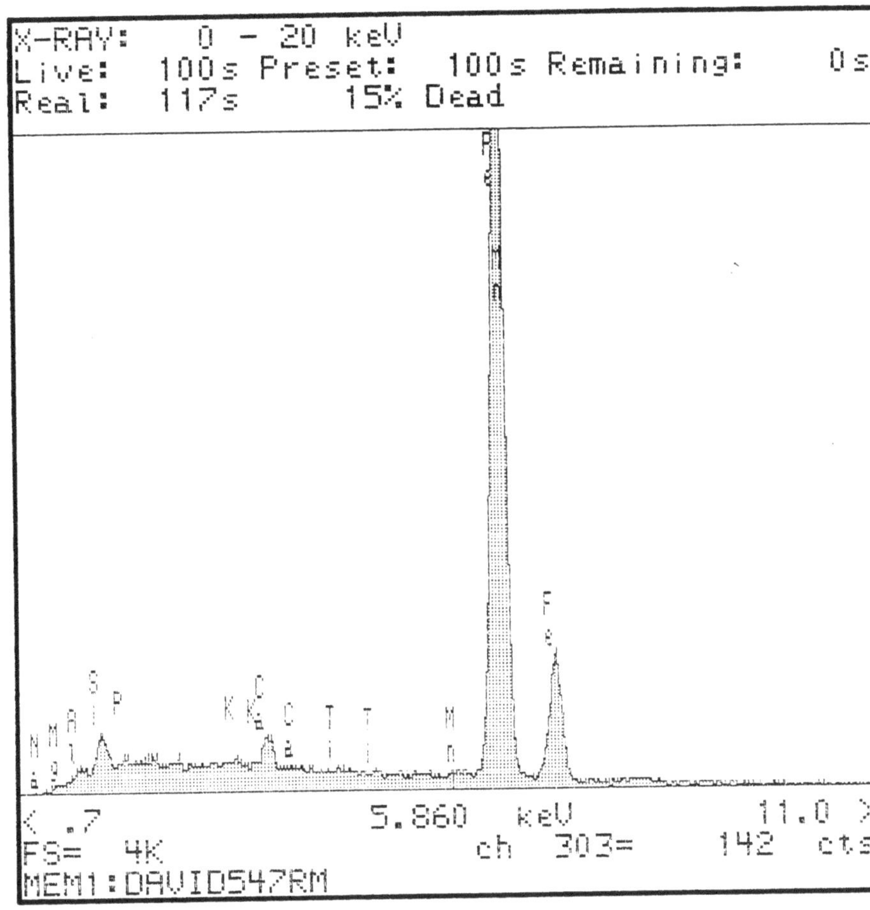

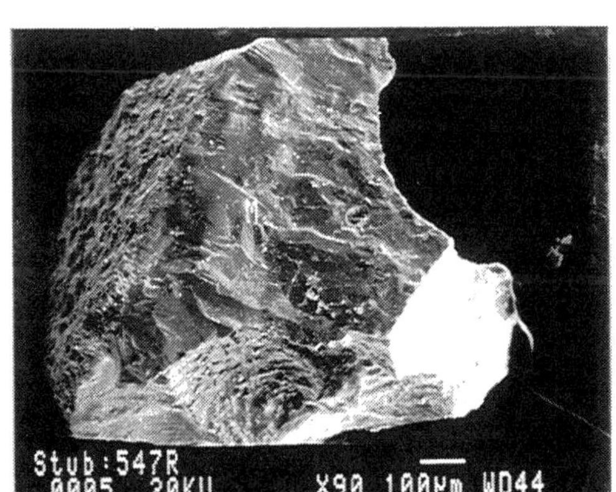

Figure 5.10 Black slag from Experiment No.2 (bloom iron forged by traditional methods).

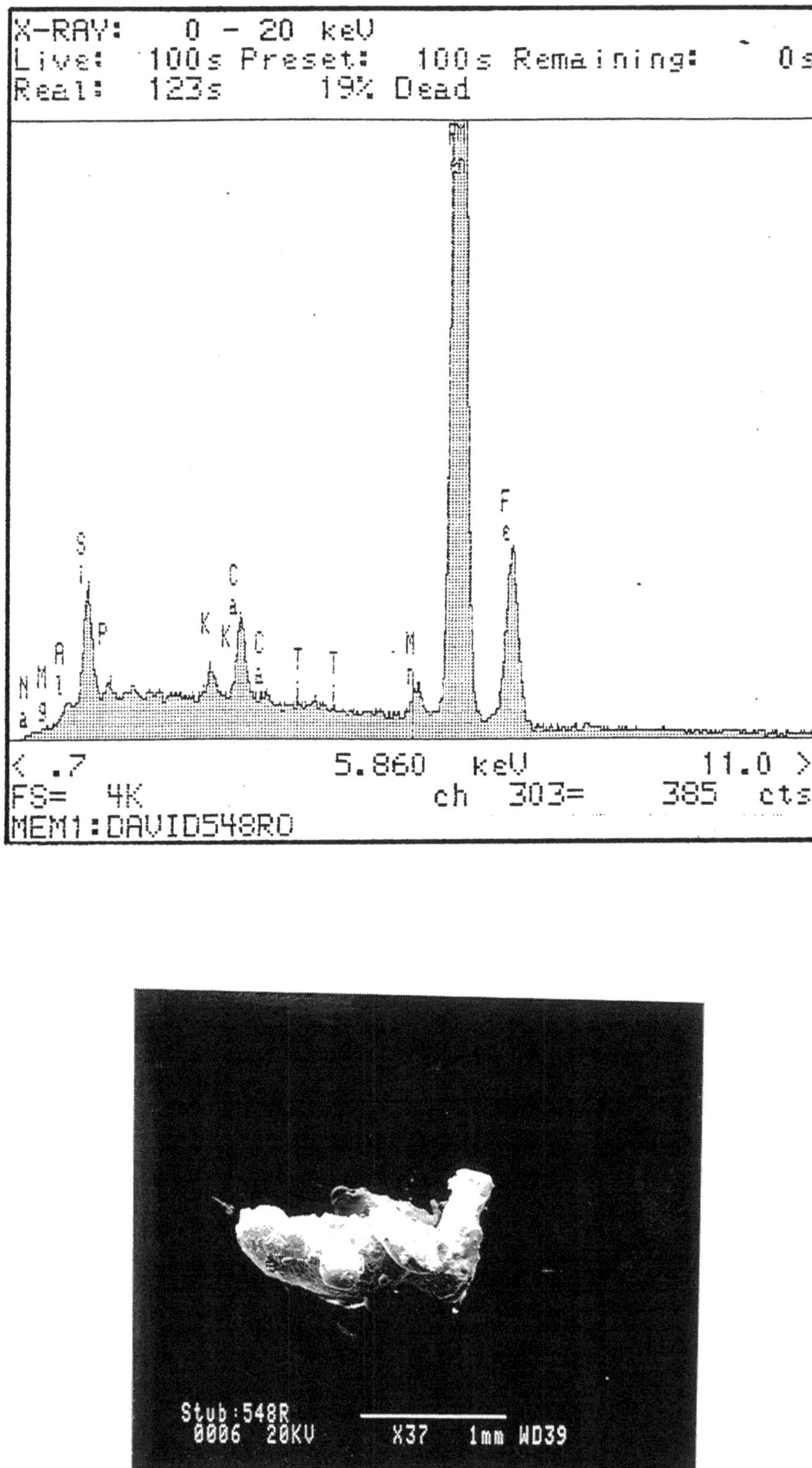

Figure 5.11 Black glass from welding bloom iron in tongs.

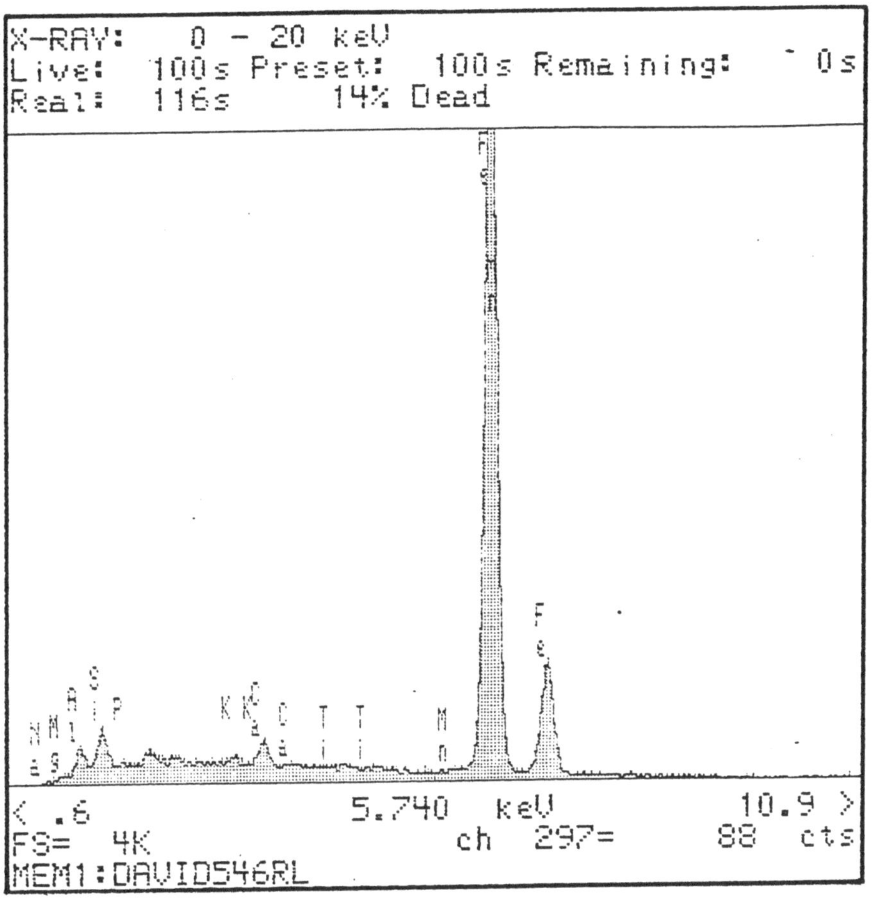

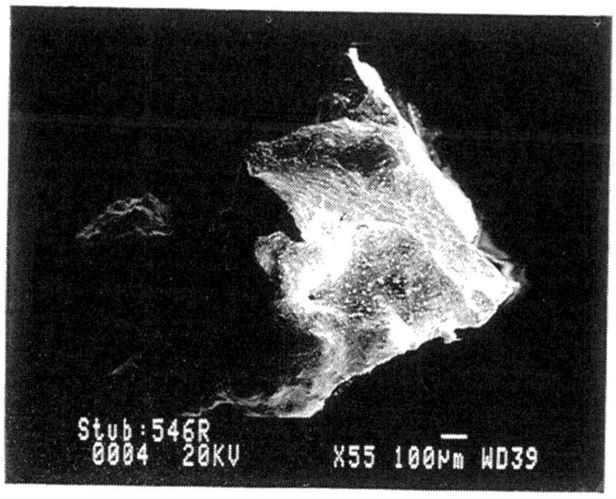

Figure 5.12 Black glass from forging iron.

Figure 5.12 shows a flake which displays the same sharp edges as other flakes and is mostly formed from iron with a small amount of silicon.

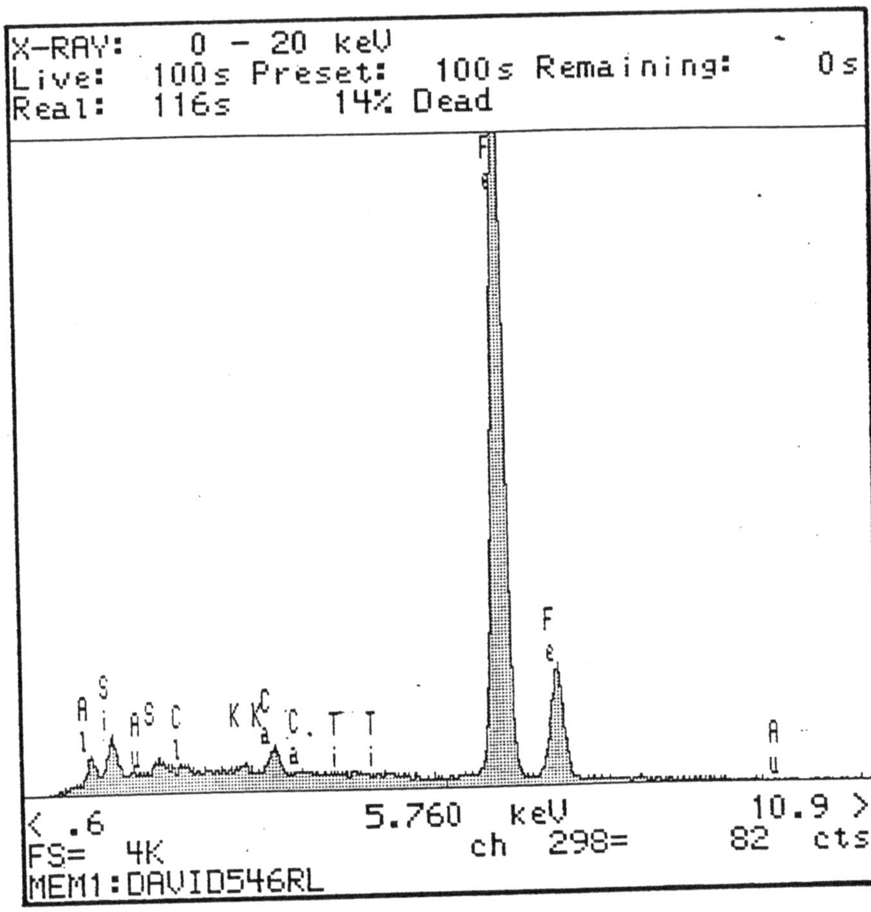

Figure 5.13 Black glass from forging iron.

Figure 5.13 shows a flake which displays the same sharp edges as other flakes and is mostly formed from iron with a small amount of silicon.

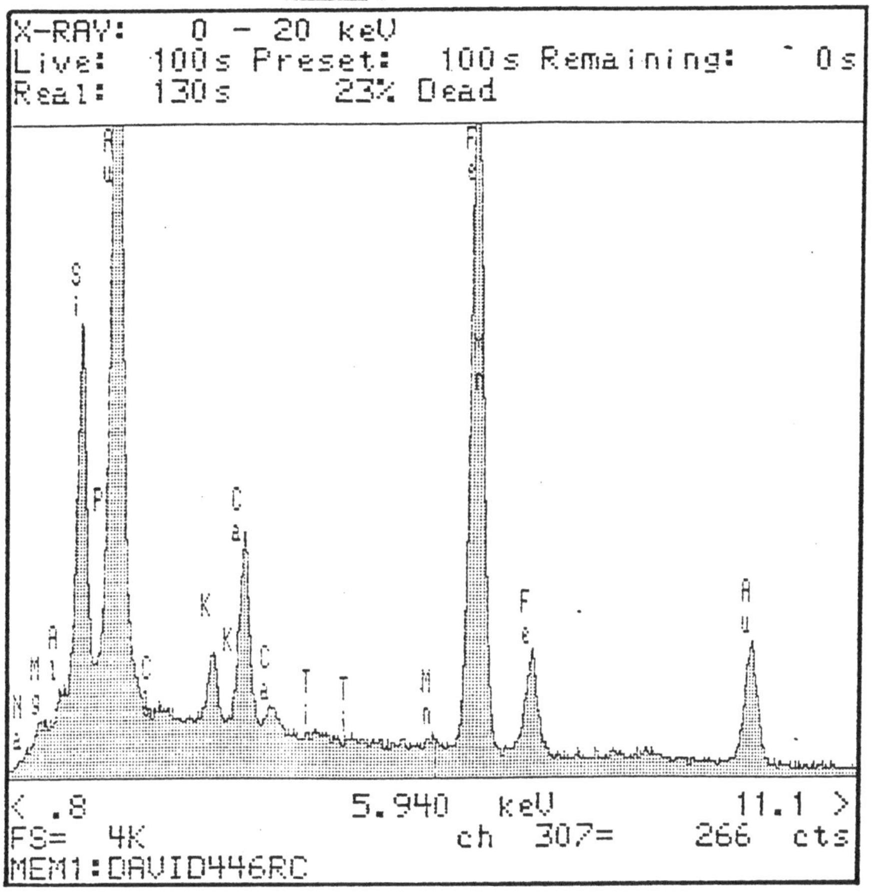

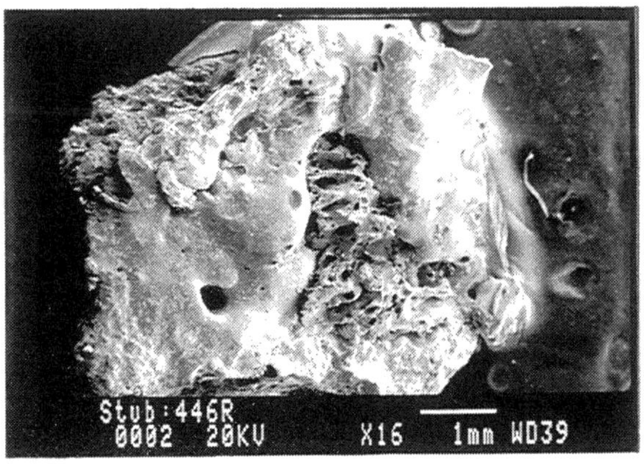

Figure 5.14 Black glass from welding a billet.

Figure 5.14 shows a flake from an experiment to weld together bloom iron which had been consolidated into a billet. This was black slag with white flakes incorporated into it. The analysis of the black portion showed it to be similar to the other black slag but the white flakes was silicon and calcium.

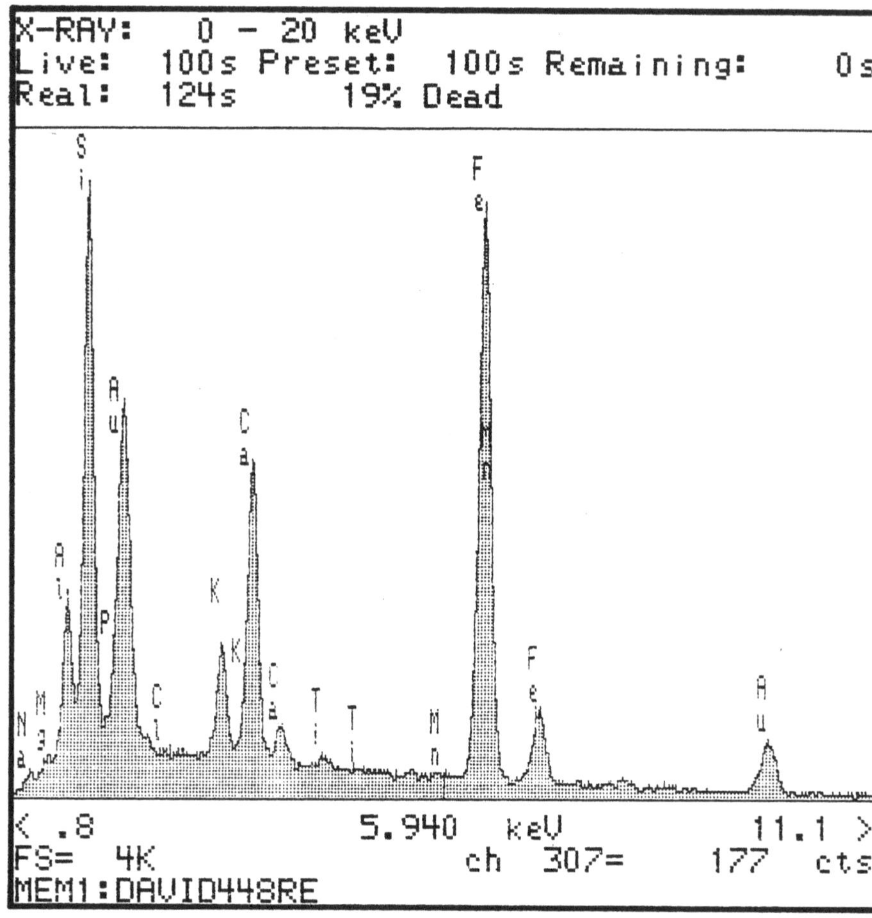

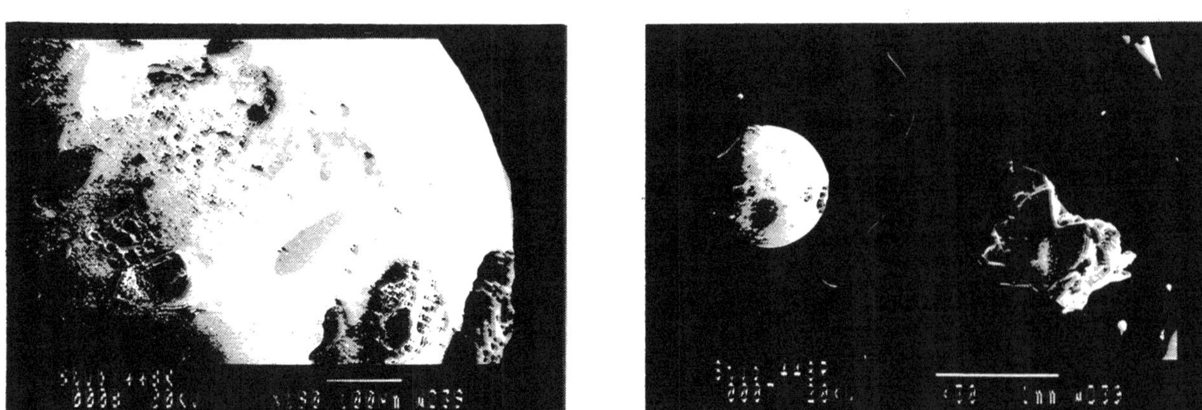

Figure 5.15 Black glass hammer scale. On the left, a sphere and on the right, a thick flake.

Figure 5.15 shows that the surface is not smooth or uniform, but has small sponge-like flakes attached to it. These appear to be attached and are not a part of the main body. The main body is not uniform but has darker patches in it.

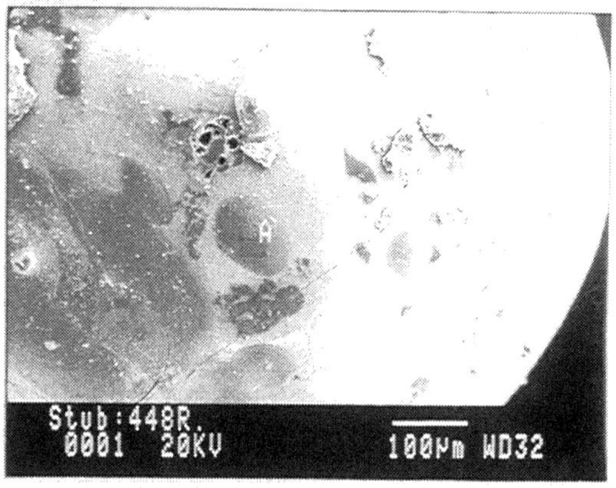

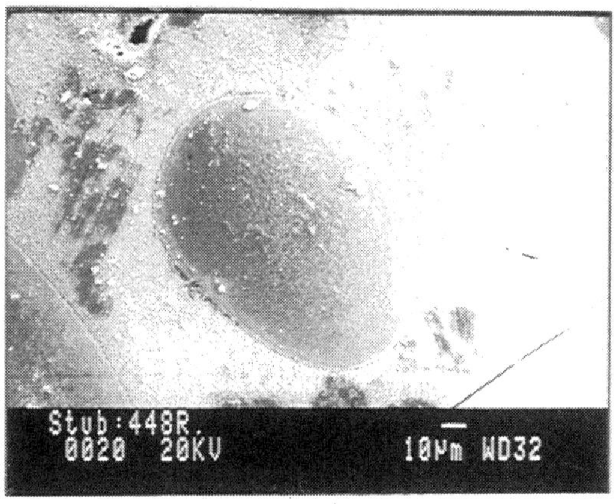

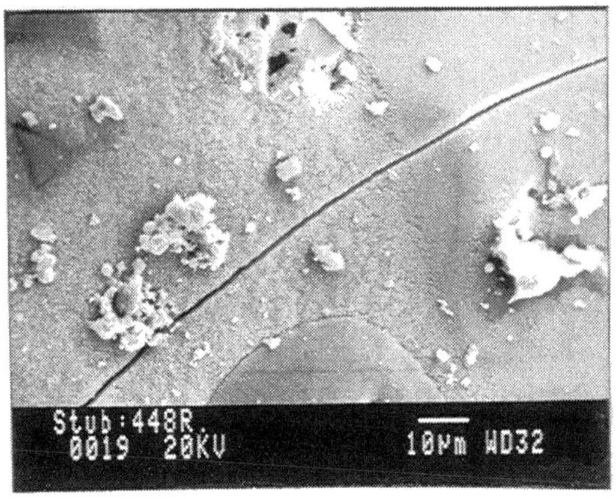

Figure 5.15 a Larger magnifications of black glass hammer scale shown in Fig 5.15.

Analysis has shown that this sample is predominantly iron and silicon in approximately even proportions, with a significant amount of calcium. The sharp cracks indicate that this is a brittle material.

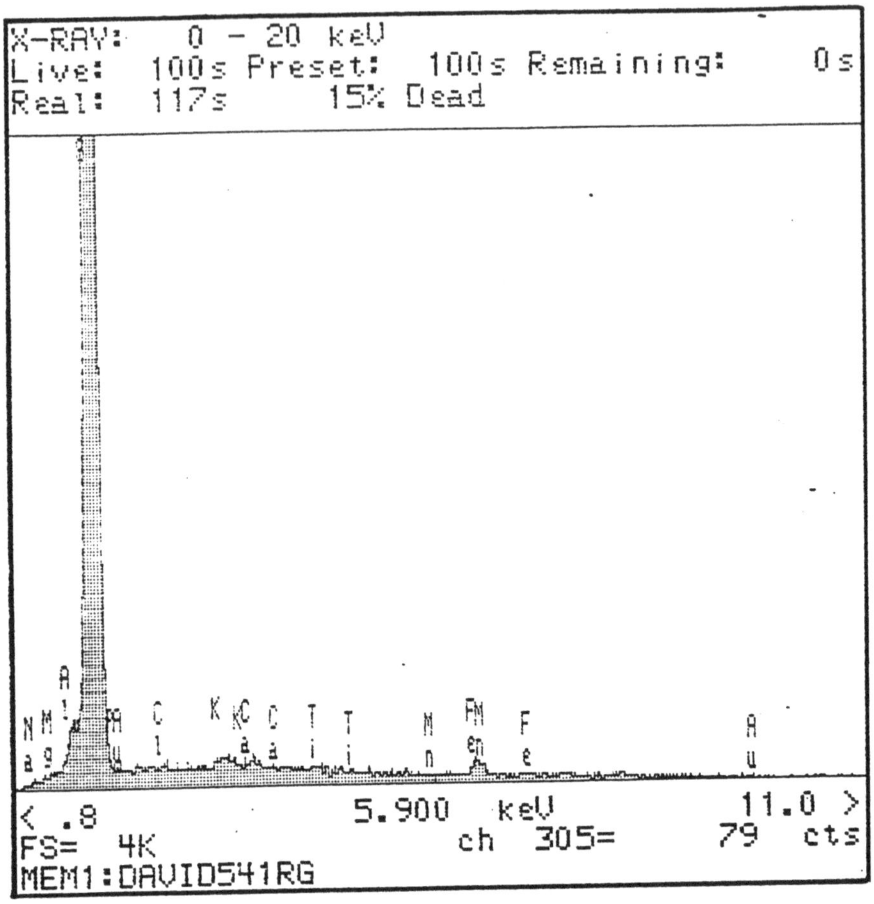

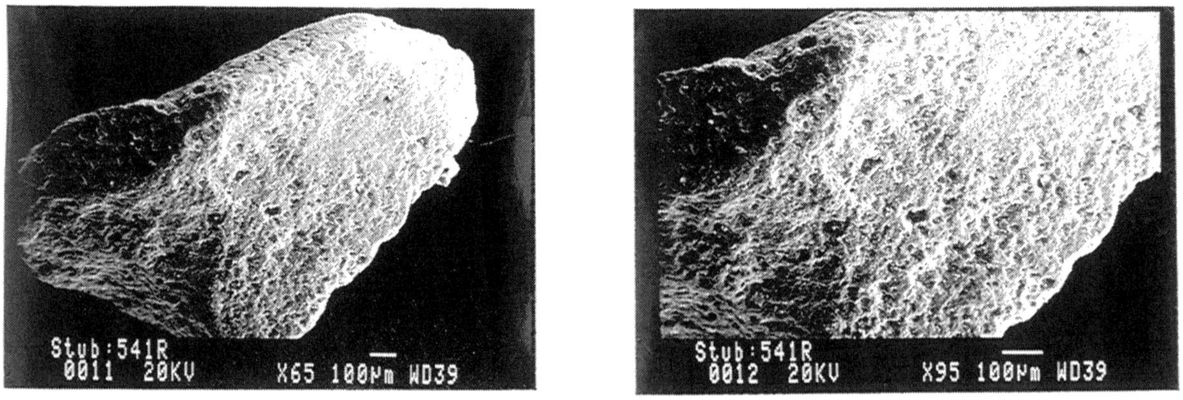

Figure 5.16 White lump from hammer scale from Silchester (almost all silicon).

This material occurs in the samples from experiments and also from the sample of hammer scale from Silchester, it is not like the flakes which are almost as thick as they are broad. As they are too small to be nuggets, the description which seemed to be appropriate was small lump.

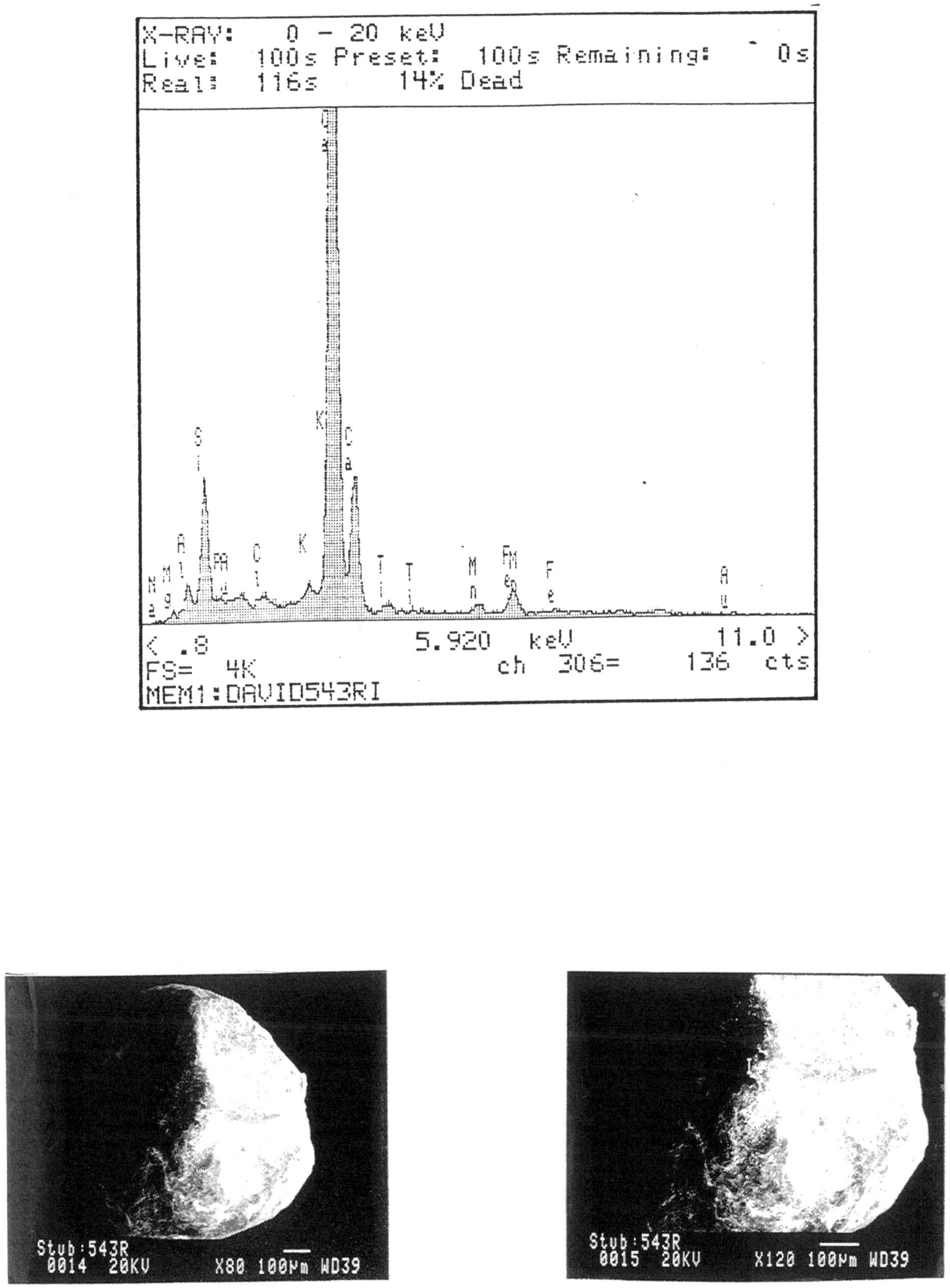

Figure 5.17 White lump from an experiment where Bloom iron was forged by traditional methods. It is mostly calcium with silicon in a ratio of about 3:1.

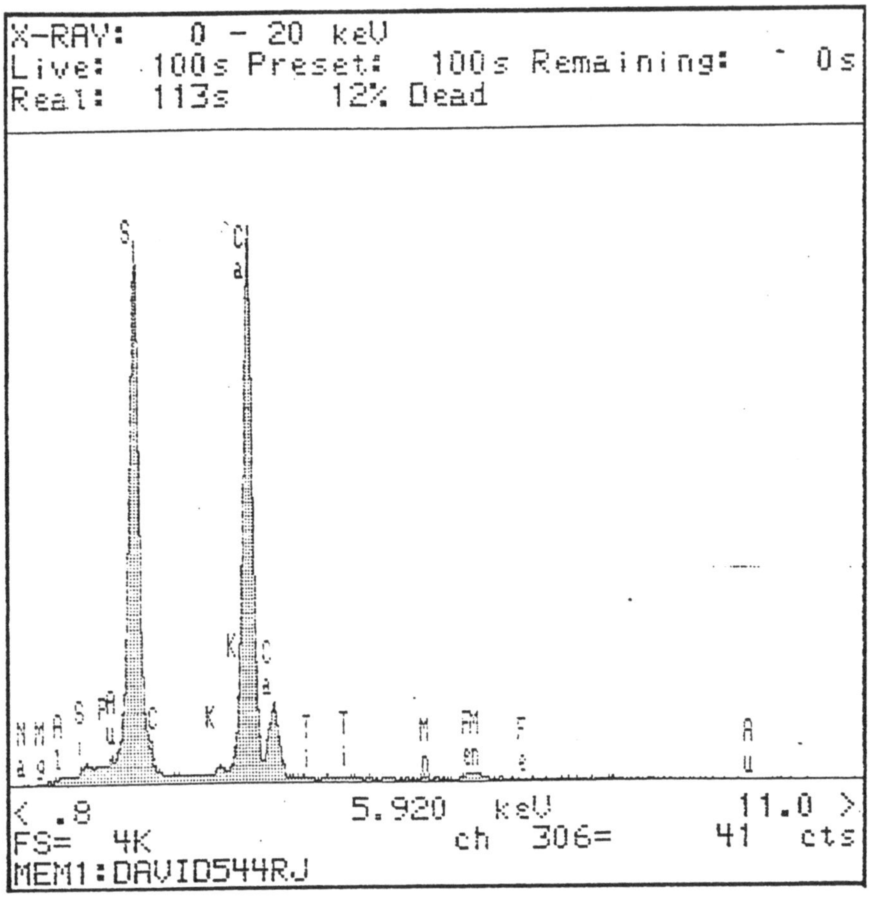

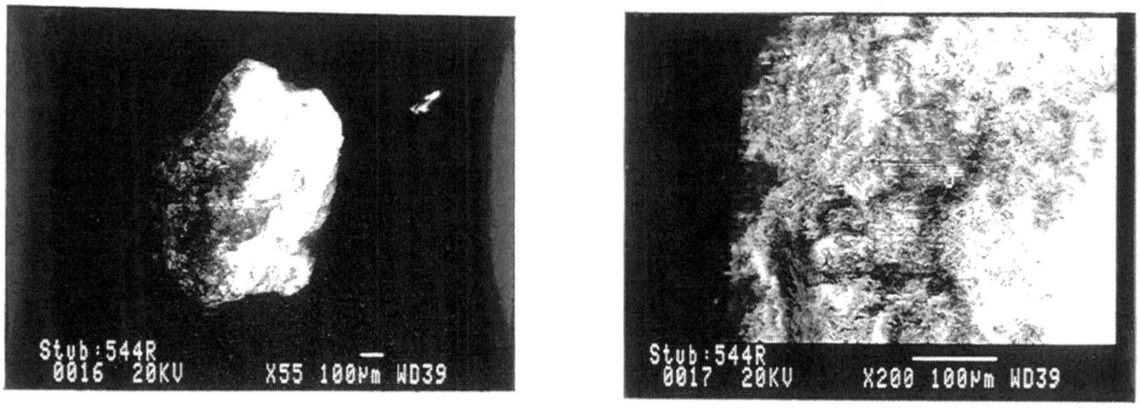

Figure 5.18 White lump from an experiment when iron was welded into a billet. It is calcium and silicon in equal amounts.

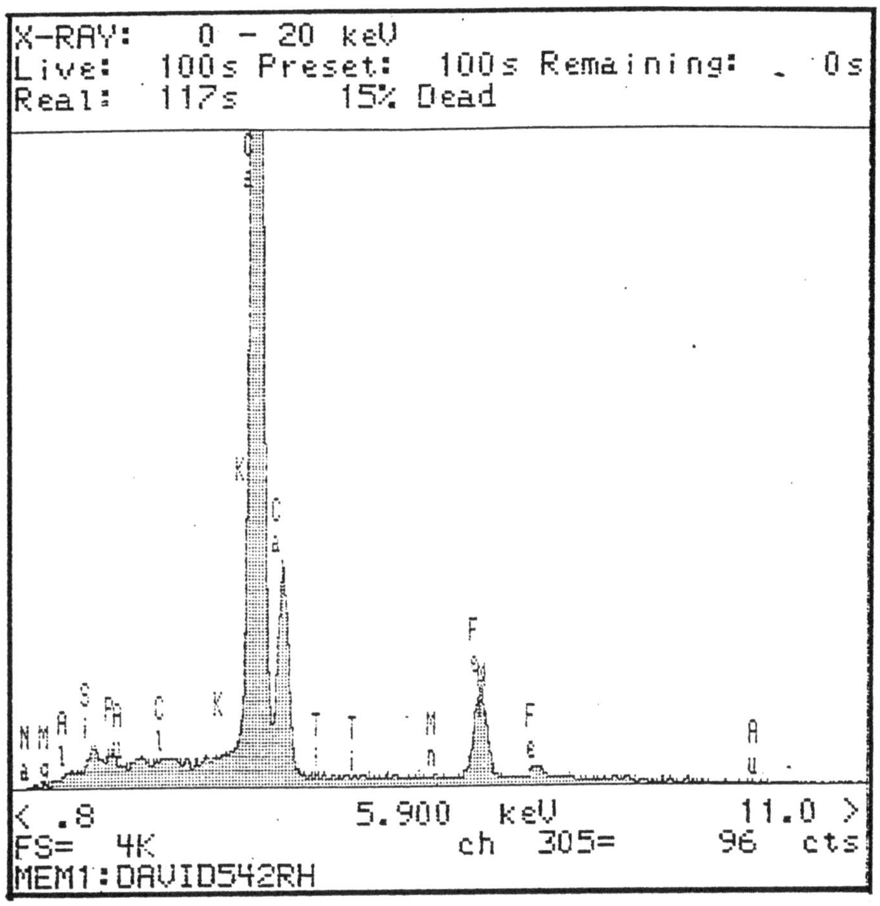

Figure 5.19 White lump from forging iron. It is mostly calcium with a small amount of silicon.

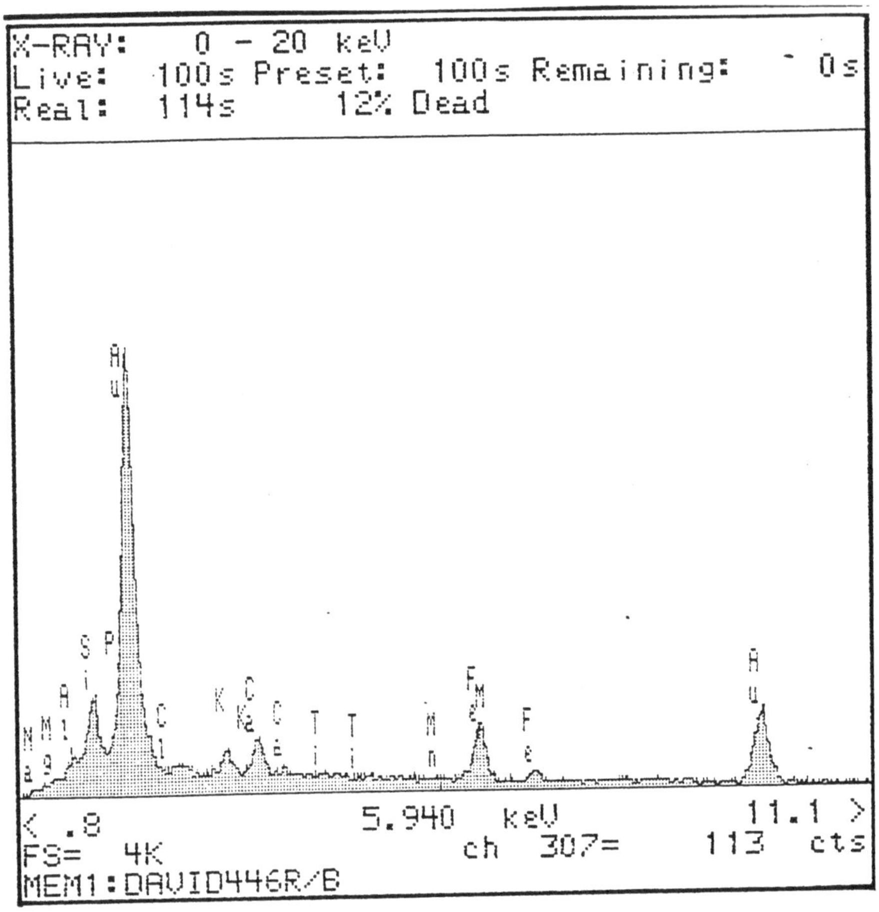

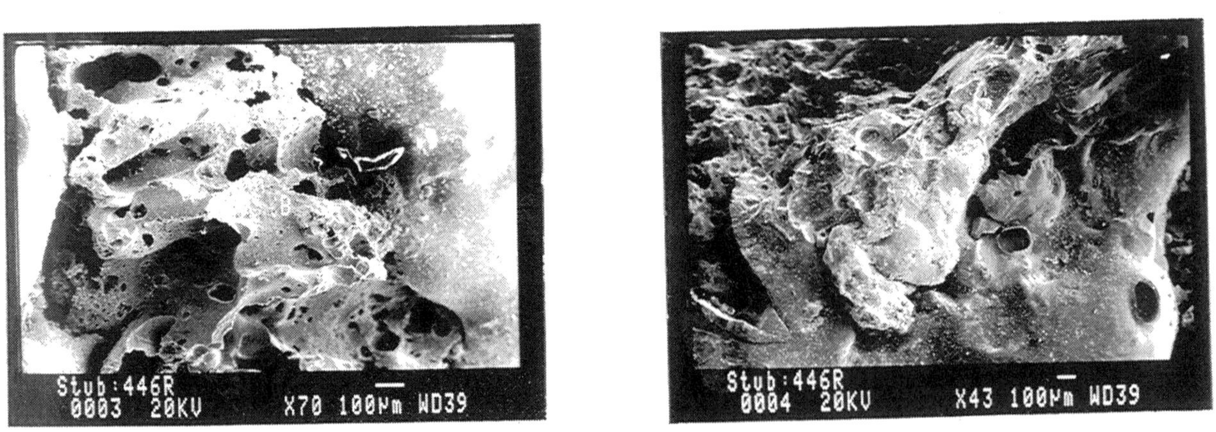

Figure 5.20 White lumps.

The samples of black glass and white nuggets found in the Basilica at Silchester are very similar to those found from debris produced in the experiments of bloomsmithing and forging of wrought iron. Since there is no evidence that points to them not coming from the same process it is probable that they were produced in the same manner.

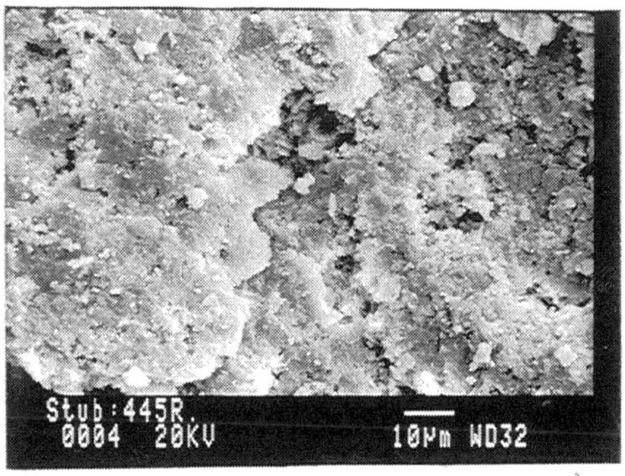

Figure 5.20 a Larger magnification of white lumps.

Experiment 1

8													
7							3.4						
6							6.9	3.4					
5						3.4	6.9	22.4					
4					3.4	6.9	6.9	3.4					
3						5.2	5.2	6.9					
2							3.4	3.4					
1								5.2	3.4				
	A	B	C	D	E	F	G	H	I	J	K	L	M

Purpose of experiment: Forging bloom iron into workable iron. Personnel: One blacksmith right handed.

Figure 5 .21 Distribution map of hammer scale (%)

Experiment 2

8					0.6	1.2	0.6	0.6					
7					1.2	1.8	0.6	1.2					
6					1.8	6.1	8.6	6.7	0.6				
5			0.6	0.6	1.2	11	17.8	2.5	0.6	0.6			
4		0.6	1.2	0.6	0.6	3.7	5.5	1.2	0.6				
3			0.6	0.6	0.6	1.2	3.7	4.3					
2						0.6	1.2	2.5					
1													
	A	B	C	D	E	F	G	H	I	J	K	L	M

Purpose of experiment: Forging bloom iron into workable iron. Personnel: One blacksmith right handed.

Figure 5.22 Distribution map of hammer scale (%)

Experiment 3

8													
7						0.3	0.3						
6						10.4	23.5	8.1					
5					0.3	4.4	**29.8**	1.3	0.3				
4					0.3	2	2.7	0.3	0.3				
3					0.3	1.7	1.3	2					
2													
1													
	A	B	C	D	E	F	G	H	I	J	K	L	M

Purpose of experiment: forging bloom iron into workable iron. Personnel: One blacksmith, right handed

Figure 5.23 Distribution map of hammer scale (%)

Experiment 4

8								1.1	1.1				
7						1.1	1.1	1.1					
6					3.3	6.6	9.9	1.1					
5					1.1	5.5	**37.4**	6.6	1.1				
4					1.1	1.1	5.5	1.1					
3						1.1		1.1					
2													
1													
	A	B	C	D	E	F	G	H	I	J	K	L	M

Purpose of experiment: forging bloom iron into workable iron. Personnel: One blacksmith, right handed

Figure 5.24 Distribution map of hammer scale (%)

Experiment 5

	A	B	C	D	E	F	G	H	I	J	K	L	M
8						0.9	0.9	0.9	0.9				
7						1.9	0.9	2.8	1.9				
6					1.9	1.9	6.6	5.7	2.8	0.9			
5				0.9	3.8	11.3	22.6	3.8	0.9	0.9			
4				0.9	0.9	3.8	2.8	1.9	0.9	0.9			
3						1.9	2.8	2.8	0.9				
2								0.9	0.9				
1								0.9	0.9				

Purpose of experiment: Welding bloom iron inside tongs. Personnel: One blacksmith right handed.

Figure 5.25 Distribution map of hammer scale (%)

Experiment 6

	A	B	C	D	E	F	G	H	I	J	K	L	M
8							1.4	1					
7					0.7	1.4	2.4	2.7	2.1				
6			0.3	0.3	0.7	2.1	8.2	11.7	3.1	1	0.7		
5			1	0.7	1	3.8	13.1	3.1	3.4	2.1	0.3		
4			0.3	1	2.1	4.8	4.5	1.7	2.7	1	0.7		
3			0.3	0.7	1	1.7	3.1	2.4	2.1	1	0.3		
2													
1													

Purpose of experiment: Forging bloom iron into workable iron. Personnel: One blacksmith right handed.

Figure 5.26 Distribution map of hammer scale (%)

Experiment 7

	A	B	C	D	E	F	G	H	I	J	K	L	M
8						1.4	1.4	1.4		1.4			
7						2	2.7	3.4	1.4				
6						3.4	4.8	4.1	1.4	1.4			
5					1.4	7.5	8.8	2.7	6.1	1.4	0.7		
4					2	6.8	7.5	2.7	5.4	2			
3					1.4	2	6.1	4.1	1.4				
2													
1													

Purpose of experiment: Forging wrought iron into bar. Personnel: One blacksmith and one striker, both right handed.

Figure 5.27 Distribution map of hammer scale (%)

Experiment 10

	A	B	C	D	E	F	G	H	I	J	K	L	M
8						0.6	0.8						
7						0.8	2.4	1.5	0.4				
6					0.4	0.8	13.5	10.5	6.6				
5					1.9	3.6	10.3	8.9	10.9				
4					0.6	4.5	5.1	2.4	1.1				
3					1.1	1.5	3.2	0.8	0.4				
2					0.4	0.8	1.5	1.1	0.6				
1						0.2	0.2	0.2					

Purpose of experiment: Forging wrought iron into bar. Personnel: One blacksmith and one striker, both right handed.

Figure 5.28 Distribution map of hammer scale (%)

Experiment 11

8						4.8	0.9	0.2	4.8				
7						1.7	1.9	0.2	0.2				
6					1.7	3.4	3.2	0.2	0.2				
5					3.9	3.6	32.1	0.2	0.2	0.2			
4				0.2	4.1	4.1	3.9	3.2	1.5	0.2			
3				0.2	0.5	3.2	3.6	3.9	0.5				
2				0.2	0.7			4.6					
1				0.2	0.7								
	A	B	C	D	E	F	G	H	I	J	K	L	M

Purpose of experiment: weld into a billet the small pieces welded in tongs on 19-10 92. Personnel: blacksmith and one striker, both right handed

Figure 5.29 Distribution map of hammer scale (%)

Experiment 13

8				0.2	1.4	0.2	0.4	0.4	0.6	0.2			
7				0.2	1.7	0.8	3.7	1.4	0.8	1.6			
6				0.6	0.8	4.1	4.1	2.1	1.7	1.2			
5				0.4	2.1	4.5	2.9	4.3	0.8	0.2			
4				0.2	3.9	5.8	17.8	2.5	2.7	3.3			
3				0.2	2.9	3.9	3.7	3.3	2.5	1.9			
2				0.2	1.7								
1													
	A	B	C	D	E	F	G	H	I	J	K	L	M

Purpose of experiment: weld into a billet the small pieces welded in tongs on 19-10 92. Personnel: blacksmith and one striker, both right handed

Figure 5.30 Distribution map of hammer scale (%)

Experiment 15

8							1.9	0.3	0.3				
7						1.6	2.8	1.6	0.3				
6						2.8	5.7	2.8	2.8				
5					0.3	6.9	13.8	15.4	1.9	2.8			
4					0.3	5.9	2.2	9.4	2.2	2.8			
3						1.6	5.3	3.2	2.8				
2													
1													
	A	B	C	D	E	F	G	H	I	J	K	L	M

Purpose of experiment: To cut a bloom with an axe then weld the pieces inside tongs.Personnel: Blacksmith and one striker, both right handed

Figure 5.31 Distribution map of hammer scale (%)

Experiment 16

8						0.3	0.7	0.3		4	0.3		
7						5	0.7	1	0.3	2.7			
6				0.3	0.3	3.4	6.4	3.4	3.4				
5				0.3	1.7	7	15.1	13.7	1.3	0.7			
4				0.3	1.3	6.7	2	1	2	0.7			
3				0.3	0.3	5	5.4	1.3	1				
2													
1													
	A	B	C	D	E	F	G	H	I	J	K	L	M

Purpose of experiment: Forging bloom iron into workable iron. Personnel: One blacksmith and one striker, both right handed.

Figure 5.32 Distribution map of hammer scale (%)

CONCLUSION

Whatever is possible does not have to happen but it will happen when it is necessary.

Anon

This book has described and discussed the processing of bloom iron into a consolidated billet, and further in the manufacture of a range of artifacts. Valuable information may be obtained form a consideration of one of the smithing by products – the hammer scale. These conclusions are not exhaustive, but summarise the main points raised in the preceding chapters.

Determination of the amount of time needed to process bloom iron into workable iron, that is iron that can be turned into artifacts

It was found that following the accepted methods of refining bloom iron a tonne of workable iron took 20,333 hours to produce. This was thought to be excessive and the method of refining unnecessarily slow and clumsy. An alternative model based on the author's own experience was tried and tested, and a time of 906.67 hours/tonne was recorded. This seems a more realistic proposition. It is recognised that it is not possible to prove that an alternative method (see chapter 3) was used, but equally there is no positive proof that the accepted method was used either.

Hearth design suitable for smithing blooms that is comparable with the archaeological record

There is evidence that some blacksmithing operations were carried out in enclosed structures that resemble a furnace rather than an open hearth. When such a structure was replicated it was found to be very efficient and the introduction of air was in such a manner that the tuyère was not in direct contact with the fuel. This overcame the problems of the tuyère blocking. It is possible that the archaeological record is not littered with the vitrified remains of tuyères because they were never there in the first place.

Production of iron artifacts using reconstruction's of Roman tools and equipment to determine production times and methods of manufacture

This set of experiments have shown that the design of some Roman weapons such as leaf-shaped spear heads, javelin heads, and arrow heads seem to be based on the principle that the minimum of metal should be used and fully functional weapons can be produced in a comparatively short time. It has also shown that although the time for each item is short when large amounts of any item are to be made (as in the case of the nails at Inchtuthil) a considerable amount of manpower and fuel is consumed in producing the finished product. Further, the making of items of a small cross-section such as the stylus (see chapter 4) require iron that has been refined to a high standard and can be considered to have possibly been expensive. Field testing of weapons (Griffiths and Sim 1993) has shown that leaving spear and javelin heads unhardened is a positive advantage, in that any damage can be easily rectified and does not require the services of a skilled metal worker. The hammer used for the experiments of bloom smithing and blacksmithing was also unhardened and the mushrooming of the head during use can easily be corrected by hammering the head back into shape when it is cold. It is suggested that many Roman iron weapons, tools and other artifacts were left unhardened deliberately and not because the Roman blacksmiths lacked the knowledge to harden them.

Study of the distribution of hammer scale to determine if the pattern is different for different processes

The distribution of hammer scale around the anvil seems to be consistent irrespective of the process being conducted. Both bloomsmithing and blacksmithing left a pattern of distribution that had the major concentration of hammer scale on the opposite side of the anvil and to the blacksmith's left. The general distribution was in a roughly circular shape.

Study the hammer scale to determine if there are; morphological differences and chemical differences for different processes

It was found that there are some morphological differences between hammer scales from different processes and that as the size of the individual components diminishes the differences become less pronounced. It was also found that as iron is refined from raw bloom iron towards workable iron the amount of slag present in the hammer scale diminishes.

Examination of the residues from contemporary forges to determine if there is any similarity between them and those produced by experimentation

The hammer scale found on the floors of the modern forges examined had the same morphology as those produced during the experiments and as the sample of hammer scale from the archaeological record. The sample from Silchester, experiments and modern forges all had four main constituents; flakes, spheres, "black glass" and "white lumps" The residues were broadly similar in morphology, and were probably produced in the same way. Some chemical variation may be accounted for by differences in raw and furnace materials.

GLOSSARY

Glossary of terms used in iron making and the craft of the blacksmith.

Annealing
Heating steel to a high temperature (red hot) and holding at this temperature for a time, followed by slow cooling. This renders the steel in the softest possible condition and relieves stress. The blacksmith's method of annealing is to keep the steel in the forge for five to seven minutes, then to bury it in ash.

Bar
A length of iron forged to a cross-section, (round, square, rectangular etc) suitable to be forged into artifacts.

Blank
A section of metal roughly shaped to the form of the finished item.

Billet
A block of purified iron with only a small quantity of slag present.

Bloom
A mass of unrefined wrought iron with large quantities of entrapped slag and voids in the structure.

Carburizing see case hardening.

Case hardening
A method of hardening the surface of a metal while keeping the interior soft. This is usually accomplished by heating the metal while it is in contact with a carbon rich material and holding it at this temperature for several hours. (see steeling or cementation)

Cast iron
Iron containing 1.8% to 4.5% carbon. It is a soft brittle material unsuitable for forging but suitable for casting.

Cementation (Steeling).
A process used for carburizing soft iron bars to make steel. The bars were heated for several days in contact with carbon.

Cold chisel
Chisel for cutting metal that is in the cold condition.

Cross peen hammer

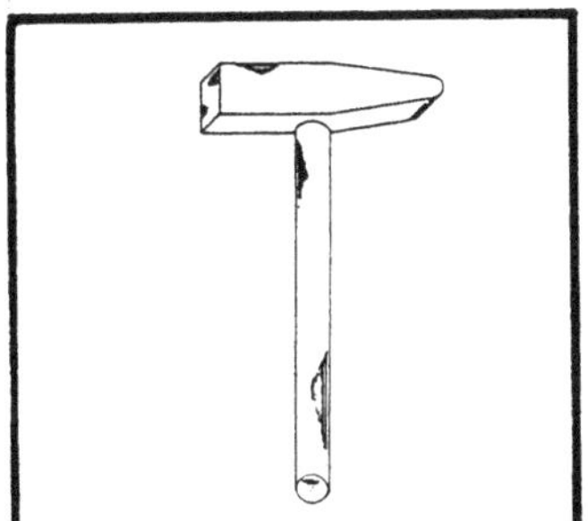

CROSS PEEN HAMMER

Drawing down
The process where the cross section of a bar is reduced by heating and hammering. The length of the bar increases correspondingly.

Drifting
A process carried out after punching to bring a hole to an accurate size and shape. The drift is a piece of hard steel forged to a taper.

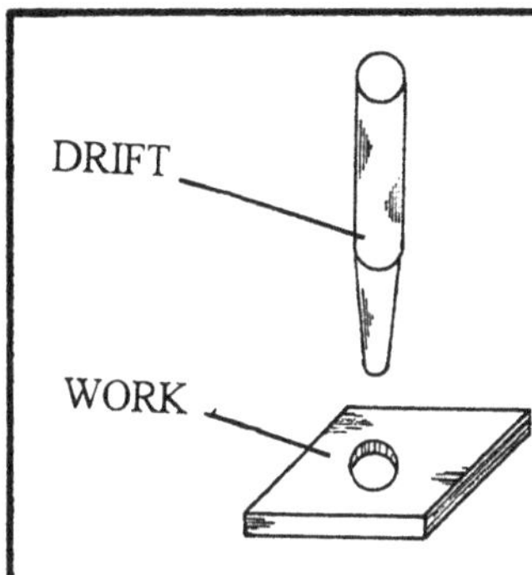

DRIFTING

Ferrite
Pure iron consists of 100% ferrite and ferrite is the principal constituent of low carbon steel.

Finery
Where the first stage of converting cast iron into wrought iron is carried out.

Flatter
Similar to a hammer but is held and struck with a sledge hammer to smooth out surface irregularities in the bar.

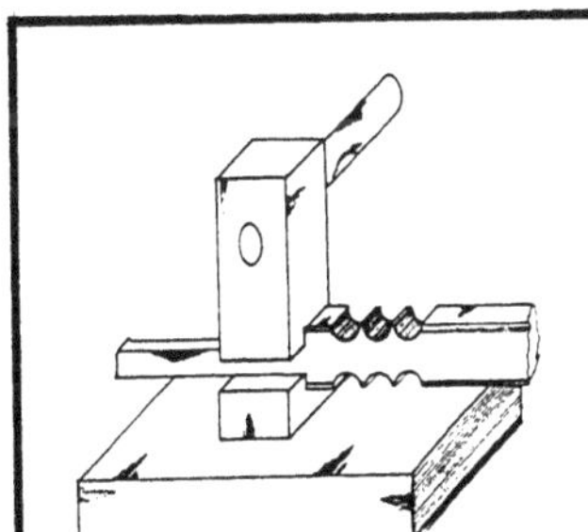

FLATTER

Flux
Substance mixed with or applied to metal surface to promote fusion.

Fire Welding (forge welding)
The process in which two pieces of iron are heated to white heat and hammered together thus causing them to fuse.

Form Tools
Made in pairs that are the reverse shape of the finished product. A blank of hot metal is placed between them and struck with a heavy force thus forming the desired shape.

Fullering
Metal is placed between two shaped tools that are struck with a hammer, this decreases the cross section and increase the length.

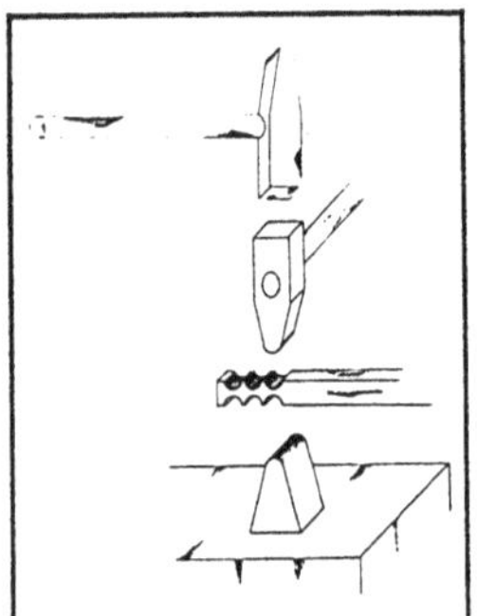

FULLERING

Gangue
Unwanted mineral.

Hardening
Steel is heated to a suitable temperature depending on carbon content; 870°C for 0.2% C to 760°C for 1.0%C and is then rapidly cooled by quenching it in water or oil. The resulting hardness will depend on the original carbon content of the steel.

Hardie Hole
Square hole in an anvil to receive tools such as bottom swages, or chisels.

Hot set
Blacksmith's chisel used for cutting metal in the hot condition (red heat and above).

Lap
A defect caused by a portion of steel being folded over on itself and failing to weld on further working.

Mandrel
Metal rod round which other metal items are forged to shape.

Punching
A punch is driven through metal to create a hole the same shape as the punch (see drifting).

Peen end of a hammer
One face of a hammer head that is shaped to perform a special task , i.e. a ball for making hemispherical indentations in the surface of a piece of metal.

Piled wrought iron
Wrought iron is rolled into strips, then cut into short lengths. The lengths are built up into a pile, placed into a furnace called a balling furnace and heated until the strips fuse. At this point it is removed and welded under a power hammer. The quality of the iron is determined by how many times this process is repeated i.e. once piled, twice piled and so on.

Plastic condition When a metal is in its most malleable condition.

Steel
An alloy of pure iron and carbon. Steels containing more than 0.3 per cent carbon can be flame hardened.

Steeling
A process by which wrought iron is converted into steel by packing the iron with carbon in a container and heating it for long periods. The carbon is absorbed into the wrought iron thus converting it to steel.

Stress Raisers
A point in a material where stresses are concentrated and where it will fracture, i.e. at a sharp corner, a sharp change of section, or a slag inclusion

Striker
Blacksmith's assistant who stands opposite the blacksmith and wields a heavy sledge hammer.

Stringers
Present in wrought iron in the form of slag. Slag starts as large inclusions and as the bloom is reduced in cross section and increases in length, the slag does the same and forms long string-like inclusions within the iron.

Swaging
Metal is worked to the desired shape by a series of blows from a hammer onto a pair of suitably shaped dies.

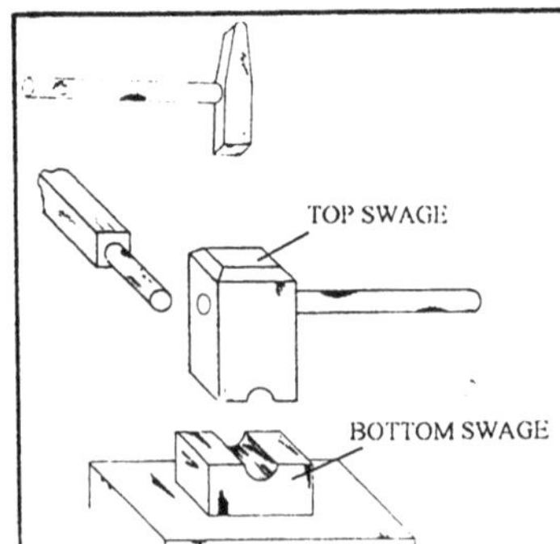

SWAGING

Tempering
Hardened steels are brittle. The brittleness is removed by heating the steel to between 210°C & 330°C then quenching in oil or water. This removes the brittleness at the expense of some of the hardness.

Upsetting
Metal is heated and forged in such a way that the length decreases and the cross section increases.

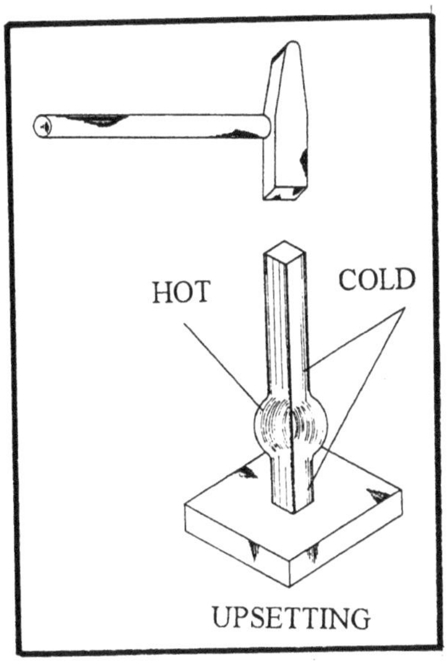

Wrought iron
The end product of the bloomery process. The structure is iron with 0.04 per cent carbon and 0.2 per cent slag. It is ductile and malleable.

BIBLIOGRAPHY

Agricola (1950) *De Re Metallica*. Trans. H.C. Hoover and L.H. Hoover. New York: Dover Publications Inc.

Aiano, R. (1975) *The Roman Iron and Steel Industries at the Time of the Empire*. Unpublished M.A. dissertation. Aberystwyth University.

Aiano, R. (1977) Romano-British Ironworking Sites, a Gazetteer. *Journal of The Historical Metallurgy Society*. 11(2), 72- 82.

Aitchison, L. (1960) *A History of Metals. Vol. 1*. London: Macdonald & Evans Ltd.

Allason-Jones,L. and Bishop,M.C.(1988) *Excavations at Roman Corbridge: the Hoard*. English Heritage Archaeological Report No 7. Historic Buildings and Monuments Commission for England

Allen, J. R. L. (1986) Interpretation of Some Romano-British Smithing Slag from Awre in Gloucestershire. *Journal of The Historical Metallurgy Society*. 20 (2), 97-104.

Allen, J. R. L. (1988) Chemical Compositional Patterns in Romano-British Bloomery Slags from the Wetlands of the Severn Estuary. *Journal of The Historical Metallurgy Society*. 22 (2), 81-86.

Angus, N. S., Brown, G. T. and Cleere, H. F. (1962) The Iron Nails from the Roman Legionary Fortress at Inchtuthil, Perthshire. *Journal of the Iron and Steel Institute*. 200 (2), 956-967.

Amman, J. (1989) *Das Standebuch*. Leipzig: Insel-Verlag.

Bachman, H. G. (1982) *The Identification of Slags from Archaeological Sites*. Occasional Paper No. 6.London: University of London Institute of Archaeology.

Biek, L. (1963) *Archaeology and the Microscope. The Scientific Examination of Archaeological Evidence*. London: Lutterworth Press.

Bishop, M. C. (1985) The Production and distribution of Roman Military Equipment. *B. A. R. International Series* 275. 1-42. Oxford: British Archeological Reports.

Bishop, M. C. and Coulston, J. C. N. (1993) *Roman Military Equipment*. London: B. T. Batsford.

Blagg, T.F.C. (1976) Tools and Techniques of the Roman stone-mason In Britain, *Brittania*,7: 165-168

Brailsford, J. W. (1962) *Hod Hill Antiquities from Hod Hill in the Durden Collection*. 1.. London: The Trustees of the British Museum.

Busch, J. A. W. (1972) Iron Making by the Bloomery Process at Nornas, Sweden, in 1851. *Journal of the Historical Metallurgy Society*. 6 (1), 28-33.

Caesar (1982) *The Conquest of Gaul*. Trans. Handford, S.A.. Middlesex: Penguin Books Ltd.

Cleere, H. F. (1971) Ironmaking in a Roman Furnace. *Britannia*. 2, 203-217.

Cleere, H. F. (1972) The Classification of Early Iron Smelting Furnaces. *The Antiquaries Journal*. 52 (1), 8-23.

Clemetson, J. (1993) Roman Scale Armour. *ARMA*. 5 (1), 8-10.

Coghlan, H. H. (1977) *Notes on Prehistoric and Early Iron in the Old World*. Occasional Papers on Technology No. 8 (2). Oxford: Pitt Rivers Museum.

Coren, S. and Porac, C. (1977) Fifty Centuries of Right-Handedness: The Historical Record. *Science*. 198, 631-632.

Crew, P. (1991) The Experimental Production of Prehistoric Bar Iron. *The Journal of the Historical Metallurgy Society*. 25 (1), 21-36.

Crew, P. (1993) Early Iron-Working Workshop, Plas Tan Bwlch. *Historical Metallurgy Society News*. 24, 1. 1-2.

Cresswell, R. G. (1991) The Radiocarbon Dating of Iron Artifacts Using Accelerator Mass Spectrometry. *Journal of The Historical Metallurgy Society*. 25 (2), 76-86.

Curle, J. (1911) *A Roman Frontier Post and its People: The Fort of Newstead in the Parish of Melrose, Glasgow*. Glasgow: James Maclehouse and Sons.

Curle, J. (1913) Notes on Some Undescribed Objects from the Roman Fort at Newstead, Melrose. *Proceedings of the Society of Antiquaries of Scotland* 47, 384-405.

Dart, R. A. (1949) The Predatory Implement Technique of Australopithecus. *American Journal of Physical Anthropology*. 7, 1-38.

David, N., Heimann, R., Killick, D. and Wayman, M. (1989) Between Bloomery and Blast Furnace: Mafa Iron-Smelting Technology in North Cameroon. *The African Archaeological Review*. 7, 183-208.

Dennis, W. (1958) Early Graphic Evidence of Dextrality in Man. *Perceptual and Motor Skills*. 8, 147-149.

Engelhard, C. (1866) *Denmark in the Early Iron Age Illustrated by the Recent Discoveries in the Peat Mosses of Slesvig*. London: Williams and Norgate.

Etienne, R. (1992) *Pompeii: The Day a City Died*. London: Thames and Hudson.

Evenstad, O. (1968) A Treatise on Iron Ore as Found in the Bogs and Swamps of Norway, and the Process of Turning it into Iron and Steel. *Bulletin of the Historical Metallurgy Group*. 2 (2), 61-65.

Frere, S. S. and Wilkes, J. J. (1989) *Strageath. Excavations within the Roman Fort,1973 to 1986. British Monograph Series No. 9*. London: Society for the Promotion of Roman Studies.

Fulford, M. G. and Allen, J. R. L. (1992) Iron-Making at the Chesters Villa, Woolaston, Gloucestershire: Survey and Excavation 1987-91. *Britannia.* 23, 159-215.

Gaitzsch, W. (1980) Eiserne Römische Werkzeuge. Vols. i and ii. *B. A. R. International Series.* 78. Oxford: British Archeological Reports.

Gilmore, B. J. J. and Tylecote, R. F. (1986) The Metallography of Early Ferrous Edge Tools and Edged Weapons. *B. A. R. British Series.* 155. Oxford: British Archeological Reports.

Gordon, R. B. (1988) Spring Strength and Structures of Wrought Iron. *Archaeomaterials.* 2 (2), 109-137.

Gralfs, B. (1988) Metalarbeitende Werkstaffen. *B. A. R. International Series.* 433. Oxford: British Archaeological Reports.

Greaves, R. and Wrighton, H. (1957) *Practical Microscopical Metallography.* London: Chapman and Hall Ltd.

Griffiths, W. and Sim, D. N. (1993) Experiments with Replica Roman Javelins. *ARBEIA Journal.* 2, 1-13.

Hannah, I. (1932) Roman Blast Furnaces in Lincolnshire. *Antiquaries Journal.* 12, 262-268.

Healy, J. F. (1978) *Mining and Metallurgy in the Greek and Roman World.* London: Thames and Hudson.

Hodge, A. T. (1990) A Roman Factory. *Scientific American.* 263 (1), 58-62.

Hodges, H. (1971) *Technology in the Ancient World.* Middlesex: Pelican Books.

James, S. (1988) The Fabricae: State Arms Factories of the Later Roman Empire. *B. A. R. International Series.* 394. Oxford: British Archeological Reports.

Jones, Alleson,L and Bishop,M. (1988*) Excavations at Roman Corbridge: The hoard.* English Heratage achaeological Report No7. Historic Buildings and Monuments Commission for England.

Jones, B., and Mattingly, D. (1990) *An Atlas of Roman Britain.* Oxford: Guild Publishing.

Jöns, H. (1993) *Eisengewinnung im norddeutschen Flachland.* In: H. Steuer and U. Zimmermann (eds). Alter Bergbau in Deutschland. Archäologie in Deutschland. Sonderheft. 63-69.

Kapp, L., Kapp, H., and Yoshindo, Y. (1987) *The Craft of the Japanese Sword.* Tokyo: Kodansha International.

Kempster, M. H. A. (1964) *Materials for Engineers.* London: The English Universities Press Ltd.

Lapsley, G. T. (1899) The Account Roll of a Fifteenth-Century Iron Master. *English Historical Review.* 14, 509-529.

Lang, J. and Ager, B. (1989) Swords of the Anglo-Saxon and Viking Periods in the British Museum: A Radiographic Study. In Hawkes, S.C., ed. *Weapons and Warfare in Anglo-Saxon England.* Monograph No. 21. Oxford: Oxford University Committee for Archaeology. 85-122.

Manning, W. H. (1972) Ironwork Hoards in Iron-age and Roman Britain. *Britannia.* 3. 224-250.

Manning, W. H. (1976) *Catalogue of Romano-British Ironwork in the Museum of Antiquities, Newcastle-upon-Tyne.* Newcastle-upon-Tyne: Department of Archaeology, University of Newcastle-upon-Tyne.

Manning, W. H.(1977) Blacksmiths' Tools from Waltham Abbey, Essex. In W.A. Oddy, ed., *Aspects of Early Metallurgy.* British Museum Research Laboratory. 1980. British Museum Occasional Papers No. 17. 87-96.

Manning, W. H. (1985) *Catalogue of the Romano-British Iron Tools, Fittings and Weapons in the British Museum.* London: British Museum Publications Ltd.

Manning, W. H. (1985) The Iron Objects. In Pitts, L., and St. Joseph, T. K., eds. *Inchtuthil The Roman Legionary Fortress.* Britannia Monograph Series 6. London; Society for the Promotion of Roman Studies. 289-299.

MacMullen, R. (1960) Inscriptions on Armour and the Supply of Arms in the Roman Empire. *American Journal of Archaeology.* 64, 23-40.

Marsden, E.W. (1971) *Greek and Roman Artillery. Technical Treatis.* Oxford. Clarendon Press. 243

Mongait, J. H. (1961) *Archaeology in the USSR.* Trans. M. W. Thompson. Harmondsworth: Penguin Books Ltd.

Muntz, A. (1972) *Die Kuntst des Metalldrehens bei Romern.*Basel

Northover, P. (1995) *Late Bronze age draw-plates in the Isleham Hord,* in B,Schmid-Sikimic and P.D.Cassa(eds),Trans Europam: Betrage zur Bronze – und Eisenzeit zwischen alantic und Altia:Festschrif fur magarita Primas,Antiquitas 3,34 , 15-22.

Ohlhaver, H. (1939) *Der Germanische Schmied und sein Werkzeug.* Leipzig: Curt Kabitzsch Verlag.

Ottaway, P. J. A. (1992) *Anglo-Scandinavian Ironwork from 16-22 Coppergate: The Small Finds.* Archaeology of York Series 17/6. London: The Council for British Archaeology.

Pietsch, M. (1984) Die Römischen Eisenwerkzeuge von Saalburg, Feldberg und Zugmantel. *Saalburg Jahrbuch.* 39, 5-132.

Pitts, L. and St Joseph, T.K. (1985) *Inchtuthil. The Roman Legionary Fortress.*Britannia Monograph SeriesNo 6. Society for the Promotion of Roman studies. 289-299

Pliny the Elder (1857) *The Natural History of Pliny. Vol. 6.* Trans. J. Bostock and H. T. Riley. London: Henry G. Bohn.

Raddaz, K. (1968) Religionsgeschichtliche Probleme des Thorsberger Moorfundes. In H.Jankuhn, ed. *Vorgeschichtliche Heiligtümer und Opferplätze in Mittel-und Nordeuropa.* Symposium, Reinhausen. Vandenhoeck and Ruprecht, Göttingen. 188-197.

Rehder, J. E. (1986) Primitive Furnaces and the Development of Metallurgy. *Journal of The Historical Metallurgy Society.* 20 (2), 87-92.

Richards, D. (1980) In M.Fulford. *Iron Age and early Roman Silchester: Excavations of the Forum Basilica.* 1980-1986. Britannia Monograph forthcoming.

Robinson Russell, H. (1975). The Armour of Imperial Rome. Purnell Book Services. London.

Robertson, A., Scott, M. and Keppie, L. (1975) *Bar Hill: A Roman Fort and its Finds.* B. A. R. British Series No. 16. Oxford: British Archeological Reports.

Röschmann, J. (1963) *Vorgeschichte des Kreises Flensburg. Neumünster.* Karl Wachholtz Verlag.

Rostoker, W., Bronson, B. and Dvorak, J. R. (1989) Smelting to Steel by the Japanese Tatara Process. *Archaeomaterials.* 3 (1), 11-25.

Scheel, B. (1989) *Egyptian Metalworking and Tools.* Aylesbury: Shire Publications Ltd.

Schubert, H. R. (1957) *History of the British Iron and Steel Industry from c. 450 B.C. to A.D. 1775.* London: Routledge and Kegan Paul.

Schulten, A. (1911) Der Ursprung des Pilums. *Rheinisches Museum 66,* 573-84.

Schulten, A. (1914) Das Pilum der Polybius. *Rheinisches Museum 69,* 477-91.

Senn, M. and Fasnacht, W. (1991) Experimental Iron Smelting and Smithing Documented by the Department of Prehistory, University of Zurich, Switzerland. *Materialy Archeologiczne.* XXVI, 27-31.

Sim,D.N. (1994) *Beyond the Bloom: Bloom Refining and artifact production in Roman Britain.* PhD. Thesis. The University of Reading.

Sim, D.N. The Role of the enclosed hearth in ancient forging techniques. (forthcoming)

Singer, C. Holmyard, J., and Hall, A. R., (eds.) (1954) *A History of Technology.* Oxford: Clarendon Press.

Souchopova, V. and Staransky, K. (1989) Experimental Iron Smelting in the Blansko Region, Czechoslovakia. *Archaeomaterials.* 3 (2), 146-161.

Springer, S. and Deutsch, G. (1989) *Left Brain, Right Brain.* New York: W. H. Freeman and Company.

Stobart, J. C. (1987) *The Grandeur That Was Rome.* London: Sidgwick and Jackson.

Theophilus (1986) *De diversis artibus.* Trans. C. R. Dodwell. Oxford: Clarendon Press.

Thomsen, E. G. and Thomsen H. H. (1974) Early Wire Drawing Through Dies. *Transactions of the ASME: Journal of Engineering for Industry.* Nov. Series B. 96 (1), 1216-1221.

Thomsen, E. G. and Thomsen H. H. (1976) Drawing Solid Wires Through Soft Dies in Antiquity. *Transactions of the ASME. Journal of Engineering for Industry.* Paper No.75-WA/Prod-6. 1-5

Thomsen, R., Naumann, F.K. and Pliner, R. (1971) *Metallographische Untersuchungen an wikingerzeitlichen Eisenbarren aus Haithabu.* Berichte über die Ausgrabunger in Haithabu, Bericht 5. K. Schietzel (ed.). Neümunster. Karl Wachholz Verlag. 100-109.

Todd, J. A. and Charles, J. A. (1978) Ethiopian Bloomery Iron and the Significance of Iron Inclusion Analysis in Iron Studies. *Journal of The Historical Metallurgy Society.* 12 (2), 63-87.

Tylecote, R. F. (1962a) Metallurgy in Archaeology, Edward Arnold, London.

Tylecote, R. F. (1962b) Roman Shaft Furnaces in Norfolk. *Journal of the Iron and Steel Institute.* 201 (1), 19-22

Tylecote, R. F. (1965) Iron Smelting in Pre-Industrial Communities. *Journal of the Iron and Steel Institute.* 203 (1), 340-348.

Tylecote, R. F. Austin, J.N. and Wraith, A.E. (1971) The Mechanism of the Bloomery Process in Shaft Furnaces. *Journal of the Iron and Steel Institute.* 209 (1), 342-363.

Tylecote, R. F. (1976) *A History of Metallurg in the British Islesy.* London: The Metals Society.

Tylecote, R. F. (1986) *The Prehistory of Metallurgy in Europe.* London. Institute of Metals.

Tylecote, R. F. (1987) *The Early History of Metallurgy in Europe.* London: Longman.

Uhrbrock, R. S. (1973) Laterality in Art. *Journal of Aesthetics and Art Criticism.* 32, 27- 35.

Wagner, D. B. (1990) Ancient Carbonization of Iron and Steel: A comment. *Archaeomaterials.* 4 (1), 111-117.

Webster, G. (1955) A Note on the Use of Coal in Roman Britain. *The Antiquaries Journal.* 19 (35), 199-216. Wheeler, R. E. M. (1943) *Maiden Castle, Dorset.* Oxford: The Society of Antiquaries. Wynne, E J. and Tylecote, R. F. (1958) An Experimental Investigation into Primitive Iron Smelting. *Journal of the Iron and Steel Institute.* 190 , 339-348.